Methods and Techniques for Proving Inequalities

Mathematical Olympiad Series

ISSN: 1793-8570

Published

Vol. 3 Graph Theory
by Xiong Bin (East China Normal University, China) &
Zheng Zhongyi (High School Attached to Fudan University, China)
translated by Liu Ruifang, Zhai Mingqing & Lin Yuanqing
(East China Normal University, China)

Vol. 4 Combinatorial Problems in Mathematical Competitions
by Yao Zhang (Hunan Normal University, P. R. China)

Vol. 5 Selected Problems of the Vietnamese Olympiad (1962–2009)
by Le Hai Chau (Ministry of Education and Training, Vietnam)
& Le Hai Khoi (Nanyang Technology University, Singapore)

Vol. 6 Lecture Notes on Mathematical Olympiad Courses:
For Junior Section (In 2 Volumes)
by Xu Jiagu

Vol. 7 A Second Step to Mathematical Olympiad Problems
by Derek Holton (University of Otago, New Zealand &
University of Melbourne, Australia)

Vol. 8 Lecture Notes on Mathematical Olympiad Courses:
For Senior Section (In 2 Volumes)
by Xu Jiagu

Vol. 9 Mathemaitcal Olympiad in China (2009–2010)
edited by Bin Xiong (East China Normal University, China) &
Peng Yee Lee (Nanyang Technological University, Singapore)

Vol. 11 Methods and Techniques for Proving Inequalities
by Yong Su (Stanford University, USA) &
Bin Xiong (East China Normal University, China)

The complete list of the published volumes in the series can be found at
http://www.worldscientific.com/series/mos

Yong Su
Stanford University, USA

Bin Xiong
East China Normal University, China

Vol. 11 | Mathematical Olympiad Series

Methods and Techniques for Proving Inequalities

Published by

East China Normal University Press
3663 North Zhongshan Road
Shanghai 200062
China

and

World Scientific Publishing Co. Pte. Ltd.
5 Toh Tuck Link, Singapore 596224
USA office: 27 Warren Street, Suite 401-402, Hackensack, NJ 07601
UK office: 57 Shelton Street, Covent Garden, London WC2H 9HE

British Library Cataloguing-in-Publication Data
A catalogue record for this book is available from the British Library.

Mathematical Olympiad Series — Vol. 11
METHODS AND TECHNIQUES FOR PROVING INEQUALITIES

ISBN 978-981-4704-12-0
ISBN 978-981-4696-45-6 (pbk)

Printed in Singapore.

Contents

CONTENTS

Chapter 1 Basic Techniques for Proving Inequalities

Among quantities observed in real life, equal relationship is local and relative while unequal relationship is universal and absolute. The exact nature of inequality is the studying of unequal relationship among quantities.

Now, for two quantities we can always compare them, or try to show one of them is greater than the other: in other words, we need to prove an inequality. The techniques for proving an inequality varies from case to case and often require some basic inequalities such as the famous AM-GM Inequality and the Cauchy-Schwarz Inequality; other techniques even involve some more advanced algebraic rearrangements. In this chapter, we introduce some of the most basic techniques for proving inequalities.

1.1 Direct Comparison

Naturally, we have two ways to compare two quantities:

(1) Compare by subtraction: to show $A \geqslant B$, it suffices to show $A - B \geqslant 0$;

(2) Compare by division: say $B > 0$, to show $A \geqslant B$, it suffices to show $\frac{A}{B} \geqslant 1$.

When we use the above two methods to compare two quantities, usually some forms of rearrangements is required. For example, factorization, separating and combining terms are some of the most used tricks.

Eg. 1 Let a, b, c be positive real numbers, prove that

$$\frac{a^2+bc}{b+c}+\frac{b^2+ca}{c+a}+\frac{c^2+ab}{a+b}\geqslant a+b+c.$$

Proof. The LHS-RHS is

$$\begin{aligned}
&\frac{a^2+bc}{b+c}-a+\frac{b^2+ca}{c+a}-b+\frac{c^2+ab}{a+b}-c\\
=&\frac{a^2+bc-ab-ac}{b+c}+\frac{b^2+ca-bc-ba}{c+a}+\frac{c^2+ab-ca-cb}{a+b}\\
=&\frac{(a-b)(a-c)}{b+c}+\frac{(b-c)(b-a)}{c+a}+\frac{(c-a)(c-b)}{a+b}\\
=&\frac{(a^2-b^2)(a^2-c^2)+(b^2-c^2)(b^2-a^2)+(c^2-a^2)(c^2-b^2)}{(b+c)(c+a)(a+b)}\\
=&\frac{a^4+b^4+c^4-a^2b^2-b^2c^2-c^2a^2}{(b+c)(c+a)(a+b)}\\
=&\frac{(a^2-b^2)^2+(b^2-c^2)^2+(c^2-a^2)^2}{2(b+c)(c+a)(a+b)}\geqslant 0,
\end{aligned}$$

therefore the original inequality holds.

Eg. 2 For real numbers x, y, z that satisfies $xy+yz+zx=-1$, show that:

$$x^2+5y^2+8z^2\geqslant 4.$$

Proof. Because

$$\begin{aligned}
&x^2+5y^2+8z^2-4\\
=&x^2+5y^2+8z^2+4(xy+yz+zx)\\
=&(x+2y+2z)^2+(y-2z)^2\geqslant 0.
\end{aligned}$$

We have $x^2+5y^2+8z^2\geqslant 4$.

Eg. 3 Let a, b, c be positive real numbers. Prove that for any real numbers x, y, z, we have:

$$\begin{aligned}
&x^2+y^2+z^2\\
\geqslant &2\sqrt{\frac{abc}{(a+b)(b+c)(c+a)}}\left(\sqrt{\frac{a+b}{c}}xy+\sqrt{\frac{b+c}{a}}yz+\sqrt{\frac{c+a}{b}}zx\right).
\end{aligned}$$

When does the equality hold?

Hint. It is well known that

$$x^2+y^2+z^2-xy-yz-zx=\frac{1}{2}[(x-y)^2+(y-z)^2+(z-x)^2]\geqslant 0,$$

we will use similar method to prove the above problem.

Proof. $LHS\text{-}RHS = \left[\frac{b}{b+c}x^2+\frac{a}{c+a}y^2-2\sqrt{\frac{ab}{(b+c)(c+a)}}xy\right]+$ $\left[\frac{c}{c+a}y^2+\frac{b}{a+b}z^2-2\sqrt{\frac{bc}{(c+a)(a+b)}}yz\right]+\left[\frac{c}{b+c}x^2+\frac{a}{a+b}z^2-\right.$ $\left.2\sqrt{\frac{ca}{(b+c)(a+b)}}xz\right]=\sum ab\left[\frac{x}{\sqrt{a(b+c)}}-\frac{y}{\sqrt{b(c+a)}}\right]^2\geqslant 0$

(here $\sum$ represents a cyclic sum).

Hence the original inequality holds.

Eg. 4 Let a, b, c be positive real numbers. Show that:

$$a^{2a}b^{2b}c^{2c}\geqslant a^{b+c}b^{c+a}c^{a+b}.$$

Proof. Since this inequality is symmetry in a, b, c, WLOG we can assume $a\geqslant b\geqslant c$, thus:

$$\frac{a^{2a}b^{2b}c^{2c}}{a^{b+c}b^{c+a}c^{a+b}}=\left(\frac{a}{b}\right)^{a-b}\left(\frac{b}{c}\right)^{b-c}\left(\frac{a}{c}\right)^{a-c}\geqslant 1.$$

Hence

$$a^{2a}b^{2b}c^{2c}\geqslant a^{b+c}b^{c+a}c^{a+b}.$$

Note. From this problem we obtain:

$$a^{3a}b^{3b}c^{3c}\geqslant a^{a+b+c}b^{a+b+c}c^{a+b+c},\text{ or: }a^ab^bc^c\geqslant(abc)^{(a+b+c)/3}.$$

Generally, if $x_i\in\mathbf{R}^+$, $i=1, 2, \cdots, n$, we have:

$$x_1^{x_1}x_2^{x_2}\cdots x_n^{x_n}\geqslant(x_1x_2\cdots x_n)^{(x_1+x_2+\cdots+x_n)/n},$$

the proof is similar to what we have done previously.

Eg. 5 Let a, b, c be positive real numbers such that $a^2+b^2+c^2=1$. Find the minimum for

$$S = \frac{1}{a^2} + \frac{1}{b^2} + \frac{1}{c^2} - \frac{2(a^3 + b^3 + c^3)}{abc}.$$

Hint. When $a = b = c$, $S = 3$. So we guess $S \geqslant 3$.

In fact,

$$\begin{aligned} S - 3 &= \frac{1}{a^2} + \frac{1}{b^2} + \frac{1}{c^2} - 3 - \frac{2(a^3 + b^3 + c^3)}{abc} \\ &= \frac{a^2 + b^2 + c^2}{a^2} + \frac{a^2 + b^2 + c^2}{b^2} + \frac{a^2 + b^2 + c^2}{c^2} - 3 - 2\left(\frac{a^2}{bc} + \frac{b^2}{ca} + \frac{c^2}{ab}\right) \\ &= a^2\left(\frac{1}{b^2} + \frac{1}{c^2}\right) + b^2\left(\frac{1}{a^2} + \frac{1}{c^2}\right) + c^2\left(\frac{1}{a^2} + \frac{1}{b^2}\right) - 2\left(\frac{a^2}{bc} + \frac{b^2}{ca} + \frac{c^2}{ab}\right) \\ &= a^2\left(\frac{1}{b} - \frac{1}{c}\right)^2 + b^2\left(\frac{1}{c} - \frac{1}{a}\right)^2 + c^2\left(\frac{1}{a} - \frac{1}{b}\right)^2 \geqslant 0. \end{aligned}$$

Therefore, the minimum for S is 3.

Note. Don't be afraid to guess the right answer (which is quite effective in handling extremum problems)! Beside guessing the right answer, we should also guess when the equality holds and keep that in mind when we later try to prove the inequality.

1.2 Method of Magnifying and Reducing

If we get stuck proving an inequality like $A \leqslant B$ directly, we can try to find a quantity C that acts like a bridge: i.e. if we have both $A \leqslant C$ and $C \leqslant B$, then $A \leqslant B$ naturally follows. In other words, we can magnify from A to C, and magnify again from C to B (the same idea applies to reduction). The trick here is to find a suitable level of magnification or reduction.

Eg. 6 Assume n is a positive integer and a_1, a_2, $\cdots$, a_n are positive real numbers.

Show that

$$\begin{aligned} &\sum_{i=1}^{n} \frac{1}{a_i^2} + \frac{1}{(a_1 + a_2 + \cdots + a_n)^2} \\ &\geqslant \frac{n^3 + 1}{(n^2 + 2011)^2}\left(\sum_{i=1}^{n} \frac{1}{a_i} + \frac{2011}{a_1 + a_2 + \cdots + a_n}\right)^2. \end{aligned}$$

Proof. From Cauchy's Inequality,

$$\left(1+1+\cdots+1+\frac{1}{n^2}\right)\left(\frac{1}{a_1^2}+\frac{1}{a_2^2}+\cdots+\frac{1}{a_n^2}+\frac{1}{(a_1+a_2+\cdots+a_n)^2}\right)$$
$$\geqslant\left(\frac{1}{a_1}+\frac{1}{a_2}+\cdots+\frac{1}{a_n}+\frac{1}{n(a_1+a_2+\cdots+a_n)}\right)^2.$$

Hence

$$\frac{1}{a_1^2}+\frac{1}{a_2^2}+\cdots+\frac{1}{a_n^2}+\frac{1}{(a_1+a_2+\cdots+a_n)^2}$$
$$\geqslant\frac{n^2}{n^3+1}\left(\frac{1}{a_1}+\frac{1}{a_2}+\cdots+\frac{1}{a_n}+\frac{1}{n(a_1+a_2+\cdots+a_n)}\right)^2.$$

It suffices to show that

$$\frac{n^2}{n^3+1}\left(\frac{1}{a_1}+\frac{1}{a_2}+\cdots+\frac{1}{a_n}+\frac{1}{n(a_1+a_2+\cdots+a_n)}\right)^2$$
$$\geqslant\frac{n^3+1}{(n^2+2011)^2}\left(\sum_{i=1}^{n}\frac{1}{a_i}+\frac{2011}{a_1+a_2+\cdots+a_n}\right)^2,$$

which is equivalent to

$$n(n^2+2011)\left(\frac{1}{a_1}+\frac{1}{a_2}+\cdots+\frac{1}{a_n}+\frac{1}{n(a_1+a_2+\cdots+a_n)}\right)$$
$$\geqslant(n^3+1)\left[\frac{1}{a_1}+\frac{1}{a_2}+\cdots+\frac{1}{a_n}+\frac{2011}{a_1+a_2+\cdots+a_n}\right],$$

which is equivalent to

$$(n^3+2011n)\sum_{i=1}^{n}\frac{1}{a_i}+(n^2+2011)\frac{1}{\sum_{i=1}^{n}a_i}$$
$$\geqslant(n^3+1)\sum_{i=1}^{n}\frac{1}{a_i}+(2011n^3+2011)\frac{1}{\sum_{i=1}^{n}a_i},$$

which is equivalent to

$$(2011n-1)\sum_{i=1}^{n}\frac{1}{a_i}\geqslant(2011n-1)n^2\frac{1}{\sum_{i=1}^{n}a_i},$$

or
$$\sum_{i=1}^{n} a_i \sum_{i=1}^{n} \frac{1}{a_i} \geqslant n^2,$$

so the conclusion holds.

Eg. 7 Prove that for all positive real numbers a, b, c, we have:

$$\frac{1}{a^3+b^3+abc}+\frac{1}{b^3+c^3+abc}+\frac{1}{c^3+a^3+abc} \leqslant \frac{1}{abc}.$$

Proof. Since $a^3+b^3=(a+b)(a^2+b^2-ab) \leqslant (a+b)ab$, we have:

$$\frac{1}{a^3+b^3+abc} \leqslant \frac{1}{ab(a+b)+abc} = \frac{c}{abc(a+b+c)}.$$

Similarly,

$$\frac{1}{b^3+c^3+abc} \leqslant \frac{a}{abc(a+b+c)},$$
$$\frac{1}{c^3+a^3+abc} \leqslant \frac{b}{abc(a+b+c)}.$$

Adding the above three inequalities together, we obtain:

$$\frac{1}{a^3+b^3+abc}+\frac{1}{b^3+c^3+abc}+\frac{1}{c^3+a^3+abc} \leqslant \frac{1}{abc}.$$

Note. When we prove inequality with fractional parts, we seldom change the fractions to a common denominator. Instead, we can simplify the fractions by means of magnifying or reduction.

Eg. 8 Assume that $a_i \geqslant 1 (i=1, 2, \cdots, n)$. Prove that:

$$(1+a_1)(1+a_2)\cdots(1+a_n) \geqslant \frac{2^n}{n+1}(1+a_1+a_2+\cdots+a_n).$$

Hint. Observing the two sides of the above inequality, how can we manage to get 2^n out from the LHS?

Proof.

$$\begin{aligned} &(1+a_1)(1+a_2)\cdots(1+a_n) \\ &= 2^n\left(1+\frac{a_1-1}{2}\right)\left(1+\frac{a_2-1}{2}\right)\cdots\left(1+\frac{a_n-1}{2}\right). \end{aligned}$$

Since $a_i - 1 \geqslant 0$, we have:

$$
\begin{aligned}
&(1+a_1)(1+a_2)\cdots(1+a_n)\\
\geqslant\ & 2^n\left(1+\frac{a_1-1}{2}+\frac{a_2-1}{2}+\cdots+\frac{a_n-1}{2}\right)\\
\geqslant\ & 2^n\left(1+\frac{a_1-1}{n+1}+\frac{a_2-1}{n+1}+\cdots+\frac{a_n-1}{n+1}\right)\\
=\ & \frac{2^n}{n+1}[n+1+(a_1-1)+(a_2-1)+\cdots+(a_n-1)]\\
=\ & \frac{2^n}{n+1}(1+a_1+a_2+\cdots+a_n).
\end{aligned}
$$

Thus the original inequality holds.

Eg. 9 Find the maximum real number α, such that $\frac{x}{\sqrt{y^2+z^2}}+\frac{y}{\sqrt{x^2+z^2}}+\frac{z}{\sqrt{x^2+y^2}}>\alpha$ holds for all real numbers x, y, z.

Soln 1. Let $x=y$, $z\to 0$, we find *LHS* $\to 2$. As a result, $\alpha>2$ will lead to a contradiction. Hence $\alpha\leqslant 2$.

WLOG, we assume $x\leqslant y\leqslant z$, let us prove that

$$\frac{x}{\sqrt{y^2+z^2}}+\frac{y}{\sqrt{x^2+z^2}}+\frac{z}{\sqrt{x^2+y^2}}>\frac{\sqrt{x^2+y^2}}{\sqrt{x^2+z^2}}+\frac{\sqrt{x^2+z^2}}{\sqrt{x^2+y^2}}.$$

Move $\frac{y}{\sqrt{z^2+x^2}}$, $\frac{z}{\sqrt{x^2+y^2}}$ to the RHS, the above inequality is equivalent to

$$\frac{x}{\sqrt{y^2+z^2}}>\frac{\sqrt{x^2+y^2}-y}{\sqrt{x^2+z^2}}+\frac{\sqrt{x^2+z^2}-z}{\sqrt{x^2+y^2}},$$

or

$$\frac{x}{\sqrt{y^2+z^2}}>\frac{x^2}{\sqrt{x^2+z^2}\,(\sqrt{x^2+y^2}+y)}+\frac{x^2}{\sqrt{x^2+y^2}\,(\sqrt{x^2+z^2}+z)}.$$

Omitting x from both sides. Since $\sqrt{x^2+y^2}+y>2y$, $\sqrt{x^2+z^2}+z>2z$, we only need to show

$$\frac{1}{\sqrt{y^2+z^2}} \geqslant \frac{x}{2y\sqrt{x^2+z^2}} + \frac{x}{2z\sqrt{x^2+y^2}}.$$

Since $\frac{x}{\sqrt{x^2+z^2}} = \frac{1}{\sqrt{1+\left(\frac{z}{x}\right)^2}}$, $\frac{x}{\sqrt{x^2+z^2}}$ increases when x increases. Similarly, $\frac{x}{\sqrt{x^2+y^2}}$ increases when x increases. Judging by these two facts, we only need to consider the case when $x = y$.

Let $x = y$, we need to prove

$$\frac{1}{\sqrt{y^2+z^2}} \geqslant \frac{1}{2\sqrt{y^2+z^2}} + \frac{1}{2\sqrt{2}z}.$$

The above inequality is equivalent to

$$\frac{1}{2\sqrt{y^2+z^2}} \geqslant \frac{1}{2\sqrt{2}z},$$

or

$$\sqrt{2}z \geqslant \sqrt{y^2+z^2},$$

which is obvious.

Therefore,

$$\frac{x}{\sqrt{y^2+z^2}} + \frac{y}{\sqrt{x^2+z^2}} + \frac{z}{\sqrt{x^2+y^2}} > \frac{\sqrt{x^2+y^2}}{\sqrt{x^2+z^2}} + \frac{\sqrt{x^2+z^2}}{\sqrt{x^2+y^2}} \geqslant 2.$$

So $\alpha_{\max} = 2$.

Note. We can also use undetermined coefficients to solve this problem.

Soln 2. Again, we show that

$$\frac{x}{\sqrt{y^2+z^2}} + \frac{y}{\sqrt{x^2+z^2}} + \frac{z}{\sqrt{x^2+y^2}} > 2.$$

Assume

$$\frac{x}{\sqrt{y^2+z^2}} \geqslant \frac{2x^a}{x^a+y^a+z^a}, \qquad ①$$

where a is an undetermined coefficient.

Note that ① is equivalent to $(x^a + y^a + z^a)^2 \geqslant 4x^{2a-2}(y^2 + z^2)$. Since $(x^a + y^a + z^a)^2 \geqslant 4x^a(y^a + z^a)$, we only need to guarantee

$$y^a + z^a \geqslant x^{a-2}(y^2 + z^2).$$

It's easy to see that $a = 2$ fits the bill. Thus,

$$\sum \frac{x}{\sqrt{y^2 + z^2}} \geqslant \sum \frac{2x^2}{x^2 + y^2 + z^2} = 2.$$

Hence, $\alpha_{\max} = 2$.

Eg. 10 Assume non-negative real numbers $a_1, a_2, \cdots, a_n$ and $b_1, b_2, \cdots, b_n$ both satisfy the following conditions:

(1) $\sum_{i=1}^{n}(a_i + b_i) = 1$;

(2) $\sum_{i=1}^{n} i(a_i - b_i) = 0$;

(3) $\sum_{i=1}^{n} i^2(a_i + b_i) = 10$.

Show that for any $1 \leqslant k \leqslant n$, $\max\{a_k, b_k\} \leqslant \frac{10}{10 + k^2}$.

Proof. For any $1 \leqslant k \leqslant n$, we have

$$\begin{aligned}(ka_k)^2 &\leqslant \left(\sum_{i=1}^{n} ia_i\right)^2 = \left(\sum_{i=1}^{n} ib_i\right)^2 \leqslant \left(\sum_{i=1}^{n} i^2 b_i\right) \cdot \left(\sum_{i=1}^{n} b_i\right) \\ &= \left(10 - \sum_{i=1}^{n} i^2 a_i\right) \cdot \left(1 - \sum_{i=1}^{n} a_i\right) \\ &\leqslant (10 - k^2 a_k) \cdot (1 - a_k) \\ &= 10 - (10 + k^2)a_k + k^2 a_k^2,\end{aligned}$$

hence $a_k \leqslant \frac{10}{10 + k^2}$. Similarly $b_k \leqslant \frac{10}{10 + k^2}$, thus

$$\max\{a_k, b_k\} \leqslant \frac{10}{10 + k^2}.$$

Eg. 11 Positive real numbers x, y, z satisfy $xyz \geqslant 1$, show that

$$\frac{x^5 - x^2}{x^5 + y^2 + z^2} + \frac{y^5 - y^2}{y^5 + z^2 + x^2} + \frac{z^5 - z^2}{z^5 + x^2 + y^2} \geqslant 0.$$

(2015 IMO)

Proof. The original inequality is equivalent to

$$\frac{x^2 + y^2 + z^2}{x^5 + y^2 + z^2} + \frac{x^2 + y^2 + z^2}{y^5 + z^2 + x^2} + \frac{x^2 + y^2 + z^2}{z^5 + x^2 + y^2} \leqslant 3.$$

From Cauchy Inequality and the given condition that $xyz \geqslant 1$, we have

$$(x^5 + y^2 + z^2)(yz + y^2 + z^2) \geqslant (x^2(xyz)^{1/2} + y^2 + z^2)^2 \geqslant (x^2 + y^2 + z^2)^2,$$

or

$$\frac{x^2 + y^2 + z^2}{x^5 + y^2 + z^2} \leqslant \frac{yz + y^2 + z^2}{x^2 + y^2 + z^2}.$$

Similarly,

$$\frac{x^2 + y^2 + z^2}{y^5 + z^2 + x^2} \leqslant \frac{zx + z^2 + x^2}{x^2 + y^2 + z^2},$$

$$\frac{x^2 + y^2 + z^2}{z^5 + x^2 + y^2} \leqslant \frac{xy + x^2 + y^2}{x^2 + y^2 + z^2}.$$

Adding these three inequalities together and use the fact that $x^2 + y^2 + z^2 \geqslant xy + yz + zx$, we have

$$\frac{x^2 + y^2 + z^2}{x^5 + y^2 + z^2} + \frac{x^2 + y^2 + z^2}{y^5 + z^2 + x^2} + \frac{x^2 + y^2 + z^2}{z^5 + x^2 + y^2} \leqslant 2 + \frac{xy + yz + zx}{x^2 + y^2 + z^2} \leqslant 3.$$

Comment. Contestant Boreico Iurie from Moldova has won the special price for his solution outlined below.

Since

$$\frac{x^5 - x^2}{x^5 + y^2 + z^2} - \frac{x^5 - x^2}{x^3(x^2 + y^2 + z^2)} = \frac{x^2(x^3 - 1)^2(y^2 + z^2)}{x^3(x^5 + y^2 + z^2)(x^2 + y^2 + z^2)} \geqslant 0.$$

As a result,

$$\sum_{\text{cyc}} \frac{x^5 - x^2}{x^5 + y^2 + z^2} \geqslant \sum_{\text{cyc}} \frac{x^5 - x^2}{x^3(x^2 + y^2 + z^2)}$$

$$= \frac{1}{x^2+y^2+z^2} \sum_{\text{cyc}} \left(x^2 - \frac{1}{x}\right)$$

$$\geqslant \frac{1}{x^2+y^2+z^2} \sum_{\text{cyc}} (x^2 - yz) (\text{since } xyz \geqslant 1)$$

$$\geqslant 0.$$

1.3 Analyzing the inequality

To analyze an inequality, we first assume that it holds and deduce from it a series of equivalent inequalities (i. e. we require that each step is reversible) until we reach an inequality that is easier or more obvious to prove than the original. Such method is usually quite helpful in gathering proving thoughts.

Eg. 12 If x, $y \in \mathbf{R}$, $y \geqslant 0$, and $y(y+1) \leqslant (x+1)^2$. Prove that

$$y(y-1) \leqslant x^2.$$

Proof. If $0 \leqslant y \leqslant 1$, then $y(y-1) \leqslant 0 \leqslant x^2$.

Now, if $y > 1$, from the assumptions we have

$$y(y+1) \leqslant (x+1)^2,$$

$$y \leqslant \sqrt{(x+1)^2 + \frac{1}{4}} - \frac{1}{2}.$$

To prove $y(y-1) \leqslant x^2$, it suffices to show that

$$\sqrt{(x+1)^2 + \frac{1}{4}} - \frac{1}{2} \leqslant \sqrt{x^2 + \frac{1}{4}} + \frac{1}{2},$$

$$\Leftrightarrow (x+1)^2 + \frac{1}{4} \leqslant x^2 + \frac{1}{4} + 2\sqrt{x^2 + \frac{1}{4}} + 1,$$

$$\Leftrightarrow 2x \leqslant 2\sqrt{x^2 + \frac{1}{4}}.$$

The final inequality is obvious, hence the original inequality holds.

Eg. 13 Assume a, b, $c \in \mathbf{R}^+$. Show that

$$a + b + c - 3\sqrt[3]{abc} \geqslant a + b - 2\sqrt{ab}.$$

Proof. Note that

$$a+b+c-3\sqrt[3]{abc} \geqslant a+b-2\sqrt{ab}$$
$$\Leftrightarrow c+2\sqrt{ab} \geqslant 3\sqrt[3]{abc}.$$

Because

$$c+2\sqrt{ab} = c+\sqrt{ab}+\sqrt{ab} \geqslant 3\sqrt[3]{c\sqrt{ab}\sqrt{ab}} = 3\sqrt[3]{abc}.$$

Therefore,

$$a+b+c-3\sqrt[3]{abc} \geqslant a+b-2\sqrt{ab}.$$

Note. To prove an inequality, sometimes we need to alter between the analyzing method and the comprehensive method. For Eg. 13, we see from the analyzing method that $c+2\sqrt{ab} \geqslant 3\sqrt[3]{abc}$ (should be true). If we insist on continuing with the analyzing method, we might produce more complications. On the other hand, the comprehensive method leads us the solution.

Eg. 14 Assume $n \in \mathbf{N}_+$. Prove that

$$\frac{1}{n+1}\left(1+\frac{1}{3}+\cdots+\frac{1}{2n-1}\right) \geqslant \frac{1}{n}\left(\frac{1}{2}+\frac{1}{4}+\cdots+\frac{1}{2n}\right). \qquad ①$$

Proof. To prove ①, it suffices to show

$$n\left(1+\frac{1}{3}+\cdots+\frac{1}{2n-1}\right) \geqslant (n+1)\left(\frac{1}{2}+\frac{1}{4}+\cdots+\frac{1}{2n}\right). \qquad ②$$

The left hand side of ② is

$$\frac{n}{2}+\frac{n}{2}+n\left(\frac{1}{3}+\cdots+\frac{1}{2n-1}\right). \qquad ③$$

The right hand side of ② is

$$\begin{aligned} &n\left(\frac{1}{2}+\frac{1}{4}+\cdots+\frac{1}{2n}\right)+\left(\frac{1}{2}+\frac{1}{4}+\cdots+\frac{1}{2n}\right) \\ &=\frac{n}{2}+n\left(\frac{1}{4}+\cdots+\frac{1}{2n}\right)+\left(\frac{1}{2}+\frac{1}{4}+\cdots+\frac{1}{2n}\right). \end{aligned} \qquad ④$$

Compare ③ and ④, if

$$\frac{n}{2} \geqslant \frac{1}{2} + \frac{1}{4} + \cdots + \frac{1}{2n}, \quad ⑤$$

and

$$\frac{1}{3} + \frac{1}{5} + \cdots + \frac{1}{2n-1} \geqslant \frac{1}{4} + \cdots + \frac{1}{2n} \quad ⑥$$

are both true (which is quite obvious), then ② holds.

In conclusion, the original inequality ① is true.

Eg. 15 Let a, b, c be positive real numbers such that $abc = 1$. Prove that

$$(a+b)(b+c)(c+a) \geqslant 4(a+b+c-1).$$

Hint. The idea is to treat a as a parameter and regard the inequality as a quadratic equation, then we can use the discriminant of the quadratic equation to prove the inequality.

Proof. WLOG, assume $a \geqslant 1$. The original inequality is equivalent to

$$a^2(b+c) + b^2(c+a) + c^2(a+b) + 6 \geqslant 4(a+b+c), \quad ①$$

that is,

$$(a^2-1)(b+c) + b^2(c+a) + c^2(a+b) + 6 \geqslant 4a + 3(b+c).$$

Since $(a+1)(b+c) \geqslant 2\sqrt{a} \cdot 2\sqrt{bc} = 4$, if we can show

$$4(a-1) + b^2(c+a) + c^2(a+b) + 6 \geqslant 4a + 3(b+c), \quad ②$$

then ① holds.

Now, ② is equivalent to

$$2 + a(b^2+c^2) + bc(b+c) - 3(b+c) \geqslant 0,$$

so we only need to show that

$$\frac{a}{2}(b+c)^2 + (bc-3)(b+c) + 2 \geqslant 0.$$

Define $f(x) = \frac{a}{2}x^2 + (bc-3)x + 2$, then its discriminant

$$\Delta = (bc - 3)^2 - 4a.$$

It suffices to show $\Delta \leqslant 0$, which is equivalent to

$$\left(\frac{1}{a} - 3\right)^2 - 4a \leqslant 0,$$

or

$$1 - 6a + 9a^2 - 4a^3 \leqslant 0,$$

or

$$(a - 1)^2(4a - 1) \geqslant 0. \quad ③$$

Since $a \geqslant 1$, the above inequality is obviously true, hence ① follows. From the above discussion, we see that the equality holds when $a = b = c = 1$.

1.4 Method of Undetermined Coefficients

Many times, we can also introduce undetermined coefficients and solve for these coefficients to prove the inequality.

Eg. 16 Assume x, y, z are real numbers that are not all 0. Find the maximum value for $\frac{xy + 2yz}{x^2 + y^2 + z^2}$.

Hint. To find the maximum for $\frac{xy + 2yz}{x^2 + y^2 + z^2}$, we only need to show that there exists a constant c, such that

$$\frac{xy + 2yz}{x^2 + y^2 + z^2} \leqslant c, \quad ①$$

and that the equal sign holds for some x, y, z.

① can be translated to $x^2 + y^2 + z^2 \geqslant \frac{1}{c}(xy + 2yz)$. Since the right hand side has terms xy and $2yz$, we split the term y^2 in the left hand side into αy^2 and $(1 - \alpha)y^2$. Since,

$$x^2 + \alpha y^2 \geqslant 2\sqrt{\alpha} \cdot xy,$$

$$(1-\alpha)y^2+z^2 \geqslant 2\sqrt{1-\alpha}\cdot yz.$$

We want $\dfrac{2\sqrt{1-\alpha}}{2\sqrt{\alpha}}=2$, which gives $\alpha=\dfrac{1}{5}$.

Soln. Since,

$$x^2+\frac{1}{5}y^2 \geqslant \frac{2}{\sqrt{5}}\cdot xy,$$

$$\frac{4}{5}y^2+z^2 \geqslant \frac{4}{\sqrt{5}}\cdot yz.$$

We obtain

$$x^2+y^2+z^2 \geqslant \frac{2}{\sqrt{5}}(xy+2yz),$$

or

$$\frac{xy+2yz}{x^2+y^2+z^2} \leqslant \frac{\sqrt{5}}{2}.$$

The equality holds when $x=1$, $y=\sqrt{5}$, $z=2$, hence the maximum $\dfrac{\sqrt{5}}{2}$ can be reached.

Eg. 17 For $\dfrac{1}{2}\leqslant x\leqslant 1$, find the maximum value of

$$(1+x)^5(1-x)(1-2x)^2.$$

Soln. Let us consider the maximum value of $[\alpha(1+x)]^5[\beta(1-x)][\gamma(2x-1)]^2$, where α, β, γ are positive integers satisfying $5\alpha-\beta+4\gamma=0$, $\alpha(1+x)=\beta(1-x)=\gamma(2x-1)$. This implies

$$\frac{\beta-\alpha}{\beta+\alpha}=\frac{\beta+\gamma}{2\gamma+\beta}.$$

Plugging in $\beta=5\alpha+4\gamma$, we have

$$0=2(3\alpha\gamma+5\alpha^2-2\gamma^2)=2(5\alpha-2\gamma)(\alpha+\gamma).$$

Let $(\alpha,\beta,\gamma)=(2,30,5)$, from AM-GM Inequality, we obtain

$$[2(1+x)]^5[30(1-x)][5(2x-1)]^2 \leqslant \left(\frac{15}{4}\right)^8.$$

The equality is achieved when $x = \frac{7}{8}$. As a result, the maximum value for $(1+x)^5(1-x)(1-2x)^2$ is $\frac{3^7 \cdot 5^5}{2^{22}}$.

Eg. 18 (Ostrowski) Assume two sets of real numbers $a_1, a_2, \cdots, a_n$ and $b_1, b_2, \cdots, b_n$ are not scaled version of each other. Real numbers $x_1, x_2, \cdots, x_n$ satisfy:

$$a_1x_1 + a_2x_2 + \cdots + a_nx_n = 0,$$
$$b_1x_1 + b_2x_2 + \cdots + b_nx_n = 1.$$

Prove that

$$x_1^2 + x_2^2 + \cdots + x_n^2 \geqslant \frac{\sum_{i=1}^{n} a_i^2}{\sum_{i=1}^{n} a_i^2 \cdot \sum_{i=1}^{n} b_i^2 - \left(\sum_{i=1}^{n} a_ib_i\right)^2}.$$

Proof 1. Assume $\sum_{i=1}^{n} x_i^2 = \sum_{i=1}^{n} x_i^2 + \alpha \sum_{i=1}^{n} a_ix_i + \beta\left(\sum_{i=1}^{n} b_ix_i - 1\right)$, where α, β are undetermined coefficients. Thus,

$$\sum_{i=1}^{n} x_i^2 = \sum_{i=1}^{n}\left(x_i + \frac{\alpha a_i + \beta b_i}{2}\right)^2 - \sum_{i=1}^{n} \frac{(\alpha a_i + \beta b_i)^2}{4} - \beta$$
$$\geqslant -\sum_{i=1}^{n} \frac{(\alpha a_i + \beta b_i)^2}{4} - \beta.$$

For the above inequality, the equal sign holds if and only if

$$x_i = -\frac{\alpha a_i + \beta b_i}{2} \quad (i = 1, 2, \cdots, n). \qquad ①$$

Substitute ① back into $\sum_{i=1}^{n} a_ix_i = 0$ and $\sum_{i=1}^{n} b_ix_i = 1$, we have

$$-\frac{1}{2}\alpha A - \frac{1}{2}\beta C = 0,$$
$$-\frac{1}{2}\alpha C - \frac{1}{2}\beta B = 1.$$

Here, $A = \sum_{i=1}^{n} a_i^2$, $B = \sum_{i=1}^{n} b_i^2$, $C = \sum_{i=1}^{n} a_i b_i$. Therefore,

$$\alpha = \frac{2C}{AB - C^2}, \ \beta = -\frac{2A}{AB - C^2}.$$

Hence

$$\sum_{i=1}^{n} x_i^2 \geqslant - \sum_{i=1}^{n} \left(\frac{\alpha a_i + \beta b_i}{2}\right)^2 - \beta = \frac{A}{AB - C^2}.$$

Note 1. There are two more ways to prove the inequality, for reference, we will mention them as follows:

Proof 2. From the Cauchy-Schwarz Inequality, for every $t \in \mathbf{R}$, we have

$$\left[\sum_{i=1}^{n} (a_i t + b_i)^2\right] \cdot (x_1^2 + x_2^2 + \cdots + x_n^2) \geqslant \left[\sum_{i=1}^{n} (a_i t + b_i) x_i\right]^2 = 1.$$

Or

$$(x_1^2 + x_2^2 + \cdots + x_n^2)(At^2 + 2Ct + B) - 1 \geqslant 0$$

always holds true.

From $\Delta \leqslant 0$ (Δ is the discriminant with respect to t), we obtain

$$\sum_{i=1}^{n} x_i^2 \geqslant \frac{A}{AB - C^2}.$$

Proof 3. (a combination of the previous two proofs) According to the conditions, for any $\lambda \in \mathbf{R}$, we have

$$\sum_{i=1}^{n} (b_i - \lambda a_i) x_i = 1.$$

By Cauchy-Schwarz Inequality,

$$\sum_{i=1}^{n} x_i^2 \cdot \sum_{i=1}^{n} (b_i - \lambda a_i)^2 \geqslant \left[\sum_{i=1}^{n} (b_i - \lambda a_i) x_i\right]^2 = 1.$$

Hence, $\sum_{i=1}^{n} x_i^2 \geqslant \frac{1}{B + \lambda^2 A - 2\lambda C}$.

Recall that our goal is to prove

$$\sum_{i=1}^{n} x_i^2 \geqslant \frac{1}{B - \frac{C^2}{A}}.$$

So we only require that

$$\lambda^2 A - 2\lambda C \leqslant -\frac{C^2}{A},$$

or

$$\lambda^2 A^2 - 2\lambda AC + C^2 \leqslant 0.$$

Picking $\lambda = \frac{C}{A}$ satisfies the above requirement.

Note 2. From this problem one can show the Fan-Todd Theorem: Let a_k, $b_k (k = 1, 2, \cdots, n)$ be two sets of real numbers not scaled of each other. Also assume that $a_i b_k \neq a_k b_i (i \neq k)$, then

$$\frac{\sum_{k=1}^{n} a_k^2}{\sum_{k=1}^{n} a_k^2 \sum_{k=1}^{n} b_k^2 - \left(\sum_{k=1}^{n} a_k b_k\right)^2} \leqslant (C_n^2)^{-2} \cdot \sum_{k=1}^{n} \left(\sum_{i \neq k} \frac{a_k}{a_i b_k - a_k b_i}\right)^2.$$

To prove the Fan-Todd Theorem, it suffices to set $x_k = (C_n^2)^{-1} \cdot \sum_{r \neq k} \frac{a_r}{a_r b_k - a_k b_r}$, we ask the reader to check that $x_1, x_2, \cdots, x_n$ satisfies the conditions.

Eg. 19 Find the maximum m_n for the function

$$f_n(x_1, x_2, \cdots, x_n) = \frac{x_1}{(1 + x_1 + \cdots + x_n)^2} + \frac{x_2}{(1 + x_2 + \cdots + x_n)^2} + \cdots + \frac{x_n}{(1 + x_n)^2} (x_i \geqslant 0).$$

Express m_n using m_{n-1} and find $\lim_{n \to \infty} m_n$.

Hint. Every denominator in f_n is rather complicated. To proceed, we should first simplify the denominators using substitutions.

Soln. Let $a_i = \frac{1}{1 + x_i + \cdots + x_n}$, $1 \leqslant i \leqslant n$. Define $a_{n+1} = 1$.

Thus,

$$1 + x_i + x_{i+1} + \cdots + x_n = \frac{1}{a_i}.$$

Also,

$$1 + x_{i+1} + x_{i+2} + \cdots + x_n = \frac{1}{a_{i+1}}.$$

Hence

$$x_i = \frac{1}{a_i} - \frac{1}{a_{i+1}}.$$

Substituting x_i's, we have

$$\begin{aligned} f_n &= \sum_{i=1}^{n} a_i^2 \left(\frac{1}{a_i} - \frac{1}{a_{i+1}} \right) = \sum_{i=1}^{n} \left(a_i - \frac{a_i^2}{a_{i+1}} \right) \\ &= (a_1 + a_2 + \cdots + a_n) - \left(\frac{a_1^2}{a_2} + \frac{a_2^2}{a_3} + \cdots + \frac{a_n^2}{1} \right). \end{aligned}$$

To find the maximum for f_n, we construct the following inequalities:

$$\begin{cases} \dfrac{a_1^2}{a_2} + \lambda_1^2 a_2 \geqslant 2\lambda_1 a_1, \\ \dfrac{a_2^2}{a_3} + \lambda_2^2 a_3 \geqslant 2\lambda_2 a_2, \\ \cdots\cdots\cdots \\ \dfrac{a_n^2}{1} + \lambda_n^2 \geqslant 2\lambda_n a_n. \end{cases} \quad ①$$

Here $\lambda_1, \lambda_2, \cdots, \lambda_n$ are parameters, $\lambda_i \geqslant 0$.

Adding ①, if we let

$$\begin{cases} 2\lambda_1 = 1, \\ 2\lambda_2 = 1 + \lambda_1^2, \\ \cdots\cdots\cdots \\ 2\lambda_n = 1 + \lambda_{n-1}^2, \end{cases} \quad ②$$

then immediately $f_n \leqslant \lambda_n^2$.

Note that $\lambda_i \geqslant \lambda_{i-1}$, and $0 \leqslant \lambda_i \leqslant 1$. Therefore, $\lim\limits_{n\to\infty} \lambda_n$ exists. It's easy to see that the limit is 1.

1.5 Normalization

When the inequality has same order in its terms, we can assume that the variables add up to a constant k. In doing so, we simplify the inequality and at the same time enhance the known conditions, both of which help us to solve the problem.

Eg. 20 Assume a, b, c are positive real numbers, prove that

$$\frac{(2a+b+c)^2}{2a^2+(b+c)^2}+\frac{(2b+c+a)^2}{2b^2+(c+a)^2}+\frac{(2c+a+b)^2}{2c^2+(a+b)^2}\leqslant 8.$$

Proof. Since each term on the left hand side is of the same order, WLOG we assume $a+b+c=3$. It suffices to show

$$\frac{(a+3)^2}{2a^2+(3-a)^2}+\frac{(b+3)^2}{2b^2+(3-b)^2}+\frac{(c+3)^2}{2c^2+(3-c)^2}\leqslant 8.$$

Define

$$f(x)=\frac{(x+3)^2}{2x^2+(3-x)^2},\ x\in\mathbf{R}^+.$$

Then

$$\begin{aligned}f(x)&=\frac{x^2+6x+9}{3(x^2-2x+3)}\\&=\frac{1}{3}\left(1+\frac{8x+6}{x^2-2x+3}\right)=\frac{1}{3}\left(1+\frac{8x+6}{(x-1)^2+2}\right)\\&\leqslant\frac{1}{3}\left(1+\frac{8x+6}{2}\right)=\frac{1}{3}(4x+4).\end{aligned}$$

So

$$f(a)+f(b)+f(c)\leqslant\frac{1}{3}(4a+4+4b+4+4c+4)=8.$$

Eg. 21 Assume $a+b+c>0$, and $ax^2+bx+c=0$ has positive real number root. Prove that

$$4\min\{a, b, c\} \leqslant a+b+c \leqslant \frac{9}{4}\max\{a, b, c\}.$$

Proof. WLOG assume $a+b+c=1$, otherwise we can replace a, b, c by $\frac{a}{a+b+c}$, $\frac{b}{a+b+c}$, $\frac{c}{a+b+c}$ respectively. First, let us prove the statement

$$\max\{a, b, c\} \geqslant \frac{4}{9}.$$

(1) If $b \geqslant \frac{4}{9}$, then the statement already holds true.

(2) If $b < \frac{4}{9}$, since $b^2 \geqslant 4ac$, we have $ac < \frac{4}{81}$.

But $a+c=1-b > \frac{5}{9}$, so if $a<0$ or $c<0$, we will have $c > \frac{5}{9}$ or $a > \frac{5}{9}$, the statement is thus true.

If $a, c \geqslant 0$, then $\left(\frac{5}{9}-c\right) \cdot c < ac < \frac{4}{81}$. Hence, $c < \frac{1}{9}$ or $c > \frac{4}{9}$. If $c < \frac{1}{9}$, then $a > \frac{4}{9}$, the statement is true.

Next we prove the statement

$$\min\{a, b, c\} \leqslant \frac{1}{4}.$$

(1) If $a \leqslant \frac{1}{4}$, the statement is already true.

(2) If $a > \frac{1}{4}$, then $b^2 \geqslant 4ac \geqslant c$, $b+c=1-a < \frac{3}{4}$. WLOG assume $c \geqslant 0$ (otherwise the statement is obviously true). So $\sqrt{c}+c \geqslant b+c < \frac{3}{4}$, or $\left(\sqrt{c}+\frac{3}{4}\right)\left(\sqrt{c}-\frac{1}{2}\right) < 0$. So $c < \frac{1}{4}$, the statement is true.

Note. The bound we present is the best: quadratic equation $\frac{4}{9}x^2 + \frac{4}{9}x + \frac{1}{9} = 0$ demonstrates that $\frac{9}{4}$ cannot be any smaller, while equation $\frac{1}{4}x^2 + \frac{2}{4}x + \frac{1}{4}$ shows that 4 cannot be any larger.

Exercise 1

1. Let x, y, z be real numbers. Prove that

$$(x^2+y^2+z^2)[(x^2+y^2+z^2)^2-(xy+yz+zx)^2]$$
$$\geqslant (x+y+z)^2[(x^2+y^2+z^2)-(xy+yz+zx)]^2.$$

2. Let m, n be positive integers and $m > n$, prove that

$$\left(1+\frac{1}{n}\right)^n < \left(1+\frac{1}{m}\right)^m.$$

3. Assume a, b, n are positive integers bigger than 1. A_{n-1} and A_n are a-ary, B_{n-1} and B_n are b-ary. A_{n-1}, A_n, B_{n-1}, B_n are defined as:

$A_n = x_n x_{n-1} \cdots x_0$, $A_{n-1} = x_{n-1} x_{n-2} \cdots x_0$ (written in a-ary form)

$B_n = x_n x_{n-1} \cdots x_0$, $B_{n-1} = x_{n-1} x_{n-2} \cdots x_0$ (written in b-ary form)

where $x_n \neq 0$, $x_{n-1} \neq 0$. Prove that when $a > b$, we have

$$\frac{A_{n-1}}{A_n} < \frac{B_{n-1}}{B_n}.$$

4. Let a, b, c be positive real numbers, prove that

$$\frac{1}{a}+\frac{1}{b}+\frac{1}{c} \leqslant \frac{a^8+b^8+c^8}{a^3b^3c^3}.$$

5. Assume real numbers a_1, a_2, $\cdots$, a_{100} satisfy:

(1) $a_1 \geqslant a_2 \geqslant \cdots \geqslant a_{100} \geqslant 0$;

(2) $a_1 + a_2 \leqslant 100$;

(3) $a_3 + a_4 + \cdots + a_{100} \leqslant 100$.

Find the maximum for $a_1^2 + a_2^2 + \cdots + a_{100}^2$.

6. Assume $5n$ real numbers r_i, s_i, t_i, u_i, v_i are all bigger than 1 $(1 \leqslant i \leqslant n)$. Denote $R = \frac{1}{n} \cdot \sum_{i=1}^{n} r_i$, $S = \frac{1}{n} \cdot \sum_{i=1}^{n} s_i$, $T = \frac{1}{n} \cdot \sum_{i=1}^{n} t_i$,

$U = \frac{1}{n} \cdot \sum_{i=1}^{n} u_i$, $V = \frac{1}{n} \cdot \sum_{i=1}^{n} v_i$. Prove the following inequality:

$$\prod_{i=1}^{n} \frac{r_i s_i t_i u_i v_i + 1}{r_i s_i t_i u_i v_i - 1} \geqslant \left(\frac{RSTUV + 1}{RSTUV - 1}\right)^n.$$

7. Assume k, n are positive integers, $1 \leqslant k < n$. $x_1, x_2, \cdots, x_k$ are k positive real numbers whose sum equals their product. Show that

$$x_1^{n-1} + x_2^{n-1} + \cdots + x_k^{n-1} \geqslant kn.$$

8. If $a, b, c \in \mathbf{R}$, show that

$$(a^2 + ab + b^2)(b^2 + bc + c^2)(c^2 + ca + a^2) \geqslant (ab + bc + ca)^3.$$

9. Prove that for any $c > 0$, there exists positive integer n and a complex sequence $a_1, a_2, \cdots, a_n$, such that

$$c \cdot \frac{1}{2^n} \sum_{\varepsilon_1, \varepsilon_2, \cdots, \varepsilon_n} | \varepsilon_1 a_1 + \varepsilon_2 a_2 + \cdots + \varepsilon_n a_n | < \left(\sum_{j=1}^{n} | a_j |^{\frac{3}{2}}\right)^{\frac{2}{3}}$$

where $\varepsilon_j \in \{-1, 1\}$, $j = 1, 2, \cdots, n$.

10. Assume a, b are positive, $\theta \in \left(0, \frac{\pi}{2}\right)$. Find the maximum for $y = a\sqrt{\sin\theta} + b\sqrt{\cos\theta}$.

11. Assume there are n real numbers, each of them has an absolute value not greater than 2 and their cubes add up to 0. Show that their sum is not greater than $\frac{2}{3}n$.

12. Assume that $n \geqslant 3$ is an integer, real numbers $x_1, x_2, \cdots, x_n \in [-1, 1]$ satisfy the relationship $\sum_{k=1}^{n} x_k^5 = 0$. Prove that

$$\sum_{k=1}^{n} x_k \leqslant \frac{8}{15} n.$$

13. Assume that the arithmetic average of n real numbers $x_1, x_2, \cdots, x_n$ is a. Show that

$$\sum_{k=1}^{n} (x_k - a)^2 \leqslant \frac{1}{2}\left(\sum_{k=1}^{n} | x_k - a |\right)^2.$$

14. Let x, y, z be non-negative real numbers. prove that

$$x(y+z-x)^2+y(z+x-y)^2+z(x+y-z)^2 \geqslant 3xyz.$$

Point out when the equality holds.

15. Assume $n \geqslant 2$ is an integer, real numbers $a_1 \geqslant a_2 \geqslant \cdots \geqslant a_n > 0$, $b_1 \geqslant b_2 \geqslant \cdots \geqslant b_n > 0$.

Also assume $a_1 a_2 \cdots a_n = b_1 b_2 \cdots b_n$, $\sum_{1 \leqslant i < j \leqslant n} (a_i - a_j) \leqslant \sum_{1 \leqslant i < j \leqslant n} (b_i - b_j)$. Show that

$$\sum_{i=1}^{n} a_i \leqslant (n-1) \cdot \sum_{i=1}^{n} b_i.$$

16. Let x, y, z be positive real numbers. Prove that

$$(xy+yz+zx)\left[\frac{1}{(x+y)^2}+\frac{1}{(y+z)^2}+\frac{1}{(z+x)^2}\right] \geqslant \frac{9}{4}.$$

Chapter 2 Identical Transformation of the Sum

When we solve an inequality, it is often required that we perform some kind of transformations to the sum. In the following, we list some of the most important transformation formulas (we ask the reader to verify that these formulas are indeed true).

(1) $a_i a_j + b_i b_j - a_i b_j - a_j b_i = (a_i - b_i)(a_j - b_j)$;

(2) $\left(\sum_{i=1}^{n} a_i\right)^2 = \sum_{i=1}^{n} a_i^2 + 2 \sum_{1 \leqslant i < j \leqslant n} a_i a_j$;

(3) $\sum_{1 \leqslant i < j \leqslant n} (a_i - a_j)^2 = n \sum_{i=1}^{n} a_i^2 - \left(\sum_{i=1}^{n} a_i\right)^2$;

(4) $\left(\sum_{i=1}^{n} a_i\right)\left(\sum_{i=1}^{n} b_i\right) = \sum_{i=1}^{n} \sum_{j=1}^{n} a_i b_j = \sum_{i=1}^{n} \sum_{j=1}^{n} a_j a_i$;

(5) $\sum_{1 \leqslant i < j \leqslant n} a_i a_j = \sum_{i=1}^{n} \left(\sum_{j=i}^{n} a_i a_j\right) = \sum_{j=1}^{n} \left(\sum_{i=1}^{j} a_i a_j\right)$;

(6) $\sum_{i=1}^{n} \sum_{j=1}^{n} a_i b_j = \frac{1}{2} \sum_{i=1}^{n} \sum_{j=1}^{n} (a_i b_j + a_j b_i)$;

(7) $a_n - a_1 = \sum_{k=1}^{n-1} (a_{k+1} - a_k)$.

Next, let us take a look at the famous Abel transformation:

First, we assume $m, n \in \mathbf{N}_+$, $m < n$, then

$$\sum_{k=m}^{n} (A_k - A_{k-1}) b_k = A_n b_n - A_{m-1} b_m + \sum_{k=m}^{n-1} A_k (b_k - b_{k+1}). \qquad ①$$

① is known as Abel's method.

On the other hand, if in ① we let $A_0 = 0$, $A_k = \sum_{i=1}^{k} a_i \ (1 \leqslant k \leqslant n)$, we obtain:

$$\sum_{k=1}^{n} a_k b_k = b_n \sum_{k=1}^{n} a_k + \sum_{k=1}^{n-1} \left(\sum_{i=1}^{k} a_i\right)(b_k - b_{k+1}). \qquad ②$$

② is known as Abel's summation by parts.

From ②, it's easy to obtain the renowned Abel Inequality: Assume $b_1 \geqslant b_2 \geqslant \cdots \geqslant b_n > 0$, $m \leqslant \sum_{k=1}^{t} a_k \leqslant M$, $t = 1, 2, \cdots, n$, then

$$b_1 m \leqslant \sum_{k=1}^{n} a_k b_k \leqslant b_1 M. \qquad ③$$

In practice, if we find a series with sum easy to find and another series with subtraction easy to find, we can consider applying the Abel transformation.

Eg. 1 Prove the Lagrange Identity:

$$\left(\sum_{i=1}^{n} a_i^2\right) \cdot \left(\sum_{i=1}^{n} b_i^2\right) = \left(\sum_{i=1}^{n} a_i b_i\right)^2 + \sum_{1 \leqslant i < j \leqslant n} (a_i b_j - a_j b_i)^2,$$

and use this identity to prove the Cauchy-Schwarz Inequality.

Proof.

$$\begin{aligned}
&\left(\sum_{i=1}^{n} a_i^2\right) \cdot \left(\sum_{i=1}^{n} b_i^2\right) - \left(\sum_{i=1}^{n} a_i b_i\right)^2 \\
&= \sum_{i=1}^{n} \sum_{j=1}^{n} a_i^2 b_j^2 - \sum_{i=1}^{n} \sum_{j=1}^{n} a_i b_i a_j b_j \\
&= \frac{1}{2} \sum_{i=1}^{n} \sum_{j=1}^{n} (a_i^2 b_j^2 + a_j^2 b_i^2 - 2 a_i b_i a_j b_j) \\
&= \frac{1}{2} \sum_{i=1}^{n} \sum_{j=1}^{n} (a_i b_j - a_j b_i)^2 \\
&= \sum_{1 \leqslant i < j \leqslant n} (a_i b_j - a_j b_i)^2.
\end{aligned}$$

Hence, the Lagrange Identity is true.

Since $\sum_{1 \leqslant i < j \leqslant n} (a_i b_j - a_j b_i)^2 \geqslant 0$, we have:

$$\left(\sum_{i=1}^{n} a_i^2\right) \cdot \left(\sum_{i=1}^{n} b_i^2\right) \geqslant \left(\sum_{i=1}^{n} a_i b_i\right)^2.$$

Thus the Cauchy-Schwarz Inequality holds.

Eg. 2 If $p > s \geqslant r > q$, $p + q = r + s$, $a_1, a_2, \cdots, a_n > 0$, then

$$\left(\sum_{i=1}^{n} a_i^p\right)\cdot\left(\sum_{i=1}^{n} a_i^q\right) \geqslant \left(\sum_{i=1}^{n} a_i^s\right)\cdot\left(\sum_{i=1}^{n} a_i^r\right).$$

Proof.

$$\begin{aligned}
&\left(\sum_{i=1}^{n} a_i^p\right)\cdot\left(\sum_{i=1}^{n} a_i^q\right) - \left(\sum_{i=1}^{n} a_i^s\right)\cdot\left(\sum_{i=1}^{n} a_i^r\right)\\
&= \sum_{i=1}^{n}\sum_{j=1}^{n}(a_i^p a_j^q - a_i^s a_j^r)\\
&= \frac{1}{2}\sum_{i=1}^{n}\sum_{j=1}^{n}(a_i^p a_j^q - a_i^s a_j^r) + \frac{1}{2}\sum_{i=1}^{n}\sum_{j=1}^{n}(a_j^p a_i^q - a_j^s a_i^r)\\
&= \frac{1}{2}\sum_{i=1}^{n}\sum_{j=1}^{n}(a_i^p a_j^q + a_j^p a_i^q - a_i^s a_j^r - a_j^s a_i^r)\\
&= \frac{1}{2}\sum_{i=1}^{n}\sum_{j=1}^{n} a_i^q a_j^q (a_i^{p-q} + a_j^{p-q} - a_i^{s-q} a_j^{r-q} - a_j^{s-q} a_i^{r-q})\\
&= \frac{1}{2}\sum_{i=1}^{n}\sum_{j=1}^{n} a_i^q a_j^q (a_i^{p-s} - a_j^{p-s})(a_i^{s-q} - a_j^{s-q})\\
&\geqslant 0.
\end{aligned}$$

Hence the original inequality is true.

Note. Formula (6) helps to make the terms in the sum more symmetric, therefore facilitates the later factoring.

Eg. 3 If $a_i > 0$, $b_i > 0$, $a_i b_i = c_i^2 + d_i^2 (i = 1, 2, \cdots, n)$, then

$$\left(\sum_{i=1}^{n} a_i\right)\left(\sum_{i=1}^{n} b_i\right) \geqslant \left(\sum_{i=1}^{n} c_i\right)^2 + \left(\sum_{i=1}^{n} d_i\right)^2.$$

The equality holds if and only if $\frac{a_i}{a_j} = \frac{b_i}{b_j} = \frac{c_i}{c_j} = \frac{d_i}{d_j}$ $(1 \leqslant i < j \leqslant n)$.

Proof.

$$\left(\sum_{i=1}^{n} a_i\right)\left(\sum_{i=1}^{n} b_i\right) - \left(\sum_{i=1}^{n} c_i\right)^2 - \left(\sum_{i=1}^{n} d_i\right)^2$$

$$= \sum_{i=1}^{n}\sum_{j=1}^{n} a_i b_j - \sum_{i=1}^{n}\sum_{j=1}^{n} c_i c_j - \sum_{i=1}^{n}\sum_{j=1}^{n} d_i d_j$$

$$= \sum_{i=1}^{n}\sum_{j=1}^{n} (a_i b_j - c_i c_j - d_i d_j)$$

$$= \frac{1}{2}\sum_{i=1}^{n}\sum_{j=1}^{n} (a_i b_j + a_j b_i - 2c_i c_j - 2d_i d_j)$$

$$= \frac{1}{2}\sum_{i=1}^{n}\sum_{j=1}^{n} \left[\frac{a_i}{a_j}(c_j^2 + d_j^2) + \frac{a_j}{a_i}(c_i^2 + d_i^2) - 2c_i c_j - 2d_i d_j\right]$$

$$= \frac{1}{2}\sum_{i=1}^{n}\sum_{j=1}^{n} \left[\left(\sqrt{\frac{a_i}{a_j}}c_j - \sqrt{\frac{a_j}{a_i}}c_i\right)^2 + \left(\sqrt{\frac{a_i}{a_j}}d_j - \sqrt{\frac{a_j}{a_i}}d_i\right)^2\right]$$

$$\geqslant 0.$$

The equality holds, if and only if:

$$\begin{cases} \sqrt{\dfrac{a_i}{a_j}}c_j - \sqrt{\dfrac{a_j}{a_i}}c_i = 0, \\ \sqrt{\dfrac{a_i}{a_j}}d_j - \sqrt{\dfrac{a_j}{a_i}}d_i = 0. \end{cases} \quad (1 \leqslant i < j \leqslant n)$$

Or

$$a_i c_j = a_j c_i, \quad ①$$

$$a_i d_j = a_j d_i. \quad ②$$

From $①^2 + ②^2$, we have $a_i^2(c_j^2 + d_j^2) = a_j^2(c_i^2 + d_i^2)$, so

$$a_i^2 a_j b_j = a_j^2 a_i b_i.$$

Hence

$$\frac{a_i}{a_j} = \frac{b_i}{b_j}. \quad ③$$

From ①, ② and ③, we immediately know that the equality holds when

$$\frac{a_i}{a_j} = \frac{b_i}{b_j} = \frac{c_i}{c_j} = \frac{d_i}{d_j} (1 \leqslant i < j \leqslant n).$$

Note. We can also initiate our proof from the right hand side of the inequality:

Proof.

$$\begin{aligned}
&\left(\sum_{i=1}^{n} c_i\right)^2 + \left(\sum_{i=1}^{n} d_i\right)^2 \\
&= \sum_{i=1}^{n}\sum_{j=1}^{n}(c_ic_j + d_id_j) \\
&\leqslant \sum_{i=1}^{n}\sum_{j=1}^{n}\sqrt{c_i^2 + d_i^2}\cdot\sqrt{c_j^2 + d_j^2} \\
&= \sum_{i=1}^{n}\sum_{j=1}^{n}\sqrt{a_ib_i}\cdot\sqrt{a_jb_j} = \sum_{i=1}^{n}\sum_{j=1}^{n}\sqrt{a_ib_j}\cdot\sqrt{a_jb_i} \\
&\leqslant \sum_{i=1}^{n}\sum_{j=1}^{n}\frac{a_ib_j + a_jb_i}{2} = \sum_{i=1}^{n}\sum_{j=1}^{n}a_ib_j \\
&= \left(\sum_{i=1}^{n}a_i\right)\cdot\left(\sum_{i=1}^{n}b_i\right).
\end{aligned}$$

Eg. 4 A real number set $\{a_0, a_1, \cdots, a_n\}$ satisfies the following conditions:

(1) $a_0 = a_n = 0$;

(2) for $1 \leqslant k \leqslant n-1$, $a_k = c + \sum_{i=k}^{n-1} a_{i-k}(a_i + a_{i+1})$.

Prove that $c \leqslant \frac{1}{4n}$.

Proof. Denote $s_k = \sum_{i=0}^{k} a_i$, $k = 1, 2, \cdots, n$. Thus,

$$s_n = \sum_{k=0}^{n} a_k = \sum_{k=0}^{n-1} a_k = nc + \sum_{k=0}^{n-1}\sum_{i=k}^{n-1} a_{i-k}(a_i + a_{i+1}).$$

Define $a_{-1} = a_{-2} = \cdots = a_{-(n-1)} = 0$, then

$$\begin{aligned}
s_n &= nc + \sum_{k=0}^{n-1}\sum_{i=0}^{n-1} a_{i-k}(a_i + a_{i+1}) \\
&= nc + \sum_{i=0}^{n-1}\sum_{k=0}^{n-1} a_{i-k}(a_i + a_{i+1}) \\
&= nc + \sum_{i=0}^{n-1}\sum_{k=0}^{i} a_{i-k}(a_i + a_{i+1})
\end{aligned}$$

$$= nc + \sum_{i=0}^{n-1}(a_i + a_{i+1})\sum_{k=0}^{i} a_{i-k}$$

$$= nc + \sum_{i=0}^{n-1}(a_i + a_{i+1}) \cdot s_i$$

$$= nc + s_1 s_0 + (s_2 - s_0)s_1 + (s_3 - s_1)s_2 + \cdots$$

$$+ (s_{n-1} - s_{n-3})s_{n-2} + (s_n - s_{n-2})s_{n-1}$$

$$= nc + s_n s_{n-1}$$

$$= nc + s_n^2.$$

Therefore $s_n^2 - s_n + nc = 0$.

Since the determinant $\Delta = 1 - 4nc \geqslant 0$, we know $c \leqslant \frac{1}{4n}$.

Eg. 5 Assume $a_n = 1 + \frac{1}{2} + \cdots + \frac{1}{n}$, $n \in \mathbf{N}_+$. Prove that for $n \geqslant 2$, we have

$$a_n^2 > 2\left(\frac{a_2}{2} + \frac{a_3}{3} + \cdots + \frac{a_n}{n}\right).$$

Hint. It would be quite complicated to prove the inequality directly via the expression of a_n. However, if we can establish the relationship between adjacent a_n's, we might simplify the question by a huge margin.

Proof.

$$a_n^2 - a_{n-1}^2 = \left(1 + \frac{1}{2} + \cdots + \frac{1}{n}\right)^2 - \left(1 + \frac{1}{2} + \cdots + \frac{1}{n-1}\right)^2$$

$$= \frac{1}{n^2} + 2 \cdot \frac{1}{n}\left(1 + \frac{1}{2} + \cdots + \frac{1}{n-1}\right)$$

$$= \frac{1}{n^2} + \frac{2}{n}\left(a_n - \frac{1}{n}\right)$$

$$= 2 \cdot \frac{a_n}{n} - \frac{1}{n^2}.$$

Thus

$$a_n^2 - a_1^2 = 2\left(\frac{a_2}{2} + \frac{a_3}{3} + \cdots + \frac{a_n}{n}\right) - \left(\frac{1}{2^2} + \frac{1}{3^2} + \cdots + \frac{1}{n^2}\right).$$

So

$$
\begin{aligned}
a_n^2 &= 2\left(\frac{a_2}{2}+\frac{a_3}{3}+\cdots+\frac{a_n}{n}\right)+\left(1-\frac{1}{2^2}-\cdots-\frac{1}{n^2}\right)\\
&> 2\left(\frac{a_2}{2}+\frac{a_3}{3}+\cdots+\frac{a_n}{n}\right)+\left(1-\frac{1}{1\times 2}-\frac{1}{2\times 3}-\cdots-\frac{1}{(n-1)\times n}\right)\\
&= 2\left(\frac{a_2}{2}+\frac{a_3}{3}+\cdots+\frac{a_n}{n}\right)+\frac{1}{n}\\
&> 2\left(\frac{a_2}{2}+\frac{a_3}{3}+\cdots+\frac{a_n}{n}\right).
\end{aligned}
$$

Therefore, the original inequality is proved.

Note. Actually, we can also use mathematical induction to prove a slightly stronger inequality:

$$a_n^2 > 2\left(\frac{a_2}{2}+\frac{a_3}{3}+\cdots+\frac{a_n}{n}\right)+\frac{1}{n}.$$

We leave the details to our readers.

Eg. 6 Assume a series $\{a_k\}$ satisfies:

$$a_{k+1} = a_k + f(n)\cdot a_k^2,\ 0 \leqslant k \leqslant n.$$

Here $0 < a_0 < 1$, $0 < f(n) \leqslant \frac{1}{n}\left(\frac{1}{a_0}-1\right)$. Prove that

$$\frac{a_0(1+f(n))}{1+(1-a_0 p)f(n)} \leqslant a_p \leqslant \frac{a_0}{1-a_0 p f(n)}\quad (0 \leqslant p \leqslant n). \qquad ①$$

The equality holds if and only if $p = 0$.

Proof. When $p = 0$, the equal sign holds for ①.

When $p \geqslant 1$, since $a_0 > 0$, $f(n) > 0$, we have $a_p > 0\ (1 \leqslant p \leqslant n)$, thus

$$a_{k+1} = a_k + f(n)a_k^2 > a_k\ (0 \leqslant k \leqslant n).$$

Then,

$$a_{k+1} = a_k + f(n)a_k^2 < a_k + f(n)a_k a_{k+1}.$$

So

$$\frac{1}{a_k} - \frac{1}{a_{k+1}} < f(n).$$

As a result,

$$\sum_{k=0}^{p-1}\left(\frac{1}{a_k} - \frac{1}{a_{k+1}}\right) < pf(n),\ p = 1, 2, \cdots, n.$$

Hence

$$\frac{1}{a_0} - \frac{1}{a_p} < pf(n).$$

So

$$\frac{1}{a_p} > \frac{1}{a_0} - pf(n) \geqslant \frac{1}{a_0} - p \cdot \frac{1}{n}\left(\frac{1}{a_0} - 1\right) \geqslant 1,$$

which implies

$$a_p < \frac{1}{\dfrac{1}{a_0} - pf(n)} = \frac{a_0}{1 - a_0 pf(n)} \leqslant 1\ (1 \leqslant p \leqslant n).$$

On the other hand, since $a_p^2 < a_p < 1$, we obtain

$$a_{k+1} < a_k + f(n) \cdot a_k.$$

Or

$$a_k > \frac{1}{1 + f(n)} a_{k+1}.$$

Thus

$$a_{k+1} > a_k + f(n)\,\frac{a_k \cdot a_{k+1}}{1 + f(n)}.$$

In other words,

$$\frac{1}{a_k} - \frac{1}{a_{k+1}} > \frac{f(n)}{1 + f(n)}.$$

So

$$\sum_{k=0}^{p-1}\left(\frac{1}{a_k} - \frac{1}{a_{k+1}}\right) > p \cdot \frac{f(n)}{1 + f(n)}\ (1 \leqslant p \leqslant n).$$

Simplifying the above inequality, we have

$$\frac{1}{a_0}-\frac{1}{a_p}>p\cdot\frac{f(n)}{1+f(n)}.$$

Or

$$\frac{1}{a_p}<\frac{1}{a_0}-\frac{p\cdot f(n)}{1+f(n)}=\frac{1+(1-a_0 p)f(n)}{a_0(1+f(n))}.$$

Therefore,

$$a_p>\frac{a_0(1+f(n))}{1+(1-a_0 p)f(n)}\ (1\leqslant p\leqslant n).$$

Combining the two sides, we have proved the original inequality.

Eg. 7 (Kai Lai Chung's Inequality) Assume $a_1\geqslant a_2\geqslant\cdots\geqslant a_n>0$, and $\sum_{i=1}^{k}a_i\leqslant\sum_{i=1}^{k}b_i\,(1\leqslant k\leqslant n)$, then

(1) $\sum_{i=1}^{n}a_i^2\leqslant\sum_{i=1}^{n}b_i^2$;

(2) $\sum_{i=1}^{n}a_i^3\leqslant\sum_{i=1}^{n}a_ib_i^2$.

Proof. (1) From the Abel transformation formula,

$$\begin{aligned}\sum_{i=1}^{n}a_i^2&=a_n\Big(\sum_{i=1}^{n}a_i\Big)+\sum_{k=1}^{n-1}\Big(\sum_{i=1}^{k}a_i\Big)(a_k-a_{k+1})\\&\leqslant a_n\Big(\sum_{i=1}^{n}b_i\Big)+\sum_{k=1}^{n-1}\Big(\sum_{i=1}^{k}b_i\Big)(a_k-a_{k+1})\\&=\sum_{i=1}^{n}a_ib_i.\end{aligned}$$

From Cauchy-Schwarz Inequality, we have

$$\sum_{i=1}^{n}a_ib_i\leqslant\Big(\sum_{i=1}^{n}a_i^2\Big)^{\frac{1}{2}}\cdot\Big(\sum_{i=1}^{n}b_i^2\Big)^{\frac{1}{2}}.$$

Hence

$$\sum_{i=1}^{n}a_i^2\leqslant\sum_{i=1}^{n}b_i^2.$$

(2)

$$\begin{aligned}\sum_{i=1}^{n} a_i^3 &= a_n^2\left(\sum_{i=1}^{n} a_i\right)+\sum_{k=1}^{n-1}\left(\sum_{i=1}^{k} a_i\right)(a_k^2-a_{k+1}^2)\\ &\leqslant a_n^2\left(\sum_{i=1}^{n} b_i\right)+\sum_{k=1}^{n-1}\left(\sum_{i=1}^{k} b_i\right)(a_k^2-a_{k+1}^2)\\ &= \sum_{i=1}^{n} a_i^2 b_i = \sum_{i=1}^{n} a_i^{\frac{3}{2}} \cdot a_i^{\frac{1}{2}} b_i\\ &\leqslant \left(\sum_{i=1}^{n} a_i^3\right)^{\frac{1}{2}} \cdot \left(\sum_{i=1}^{n} a_i b_i^2\right)^{\frac{1}{2}}.\end{aligned}$$

Therefore

$$\sum_{i=1}^{n} a_i^3 \leqslant \sum_{i=1}^{n} a_i b_i^2.$$

Eg. 8 Assume a_1, a_2, $\cdots$ is sequence of positive real numbers satisfying the following condition: for all i, $j = 1, 2, \cdots$, $a_{i+j} \leqslant a_i + a_j$. Prove that, for positive integer n, we have:

$$a_1 + \frac{a_2}{2} + \frac{a_3}{3} + \cdots + \frac{a_n}{n} \geqslant a_n.$$

Hint. According to the condition, when $i + j = k$, we have $a_i + a_{k-i} \geqslant a_k$, from this we can estimate the sum $s_{k-1} = a_1 + \cdots + a_{k-1}$. Since it is easy to find the subtraction $\frac{1}{i} - \frac{1}{i+1}$, we would want to use Abel's transformation formula.

Proof. Denote $s_i = a_1 + a_2 + \cdots + a_i$, $i = 1, 2, \cdots, n$. Write $s_0 = 0$, then

$$2s_i = (a_1 + a_i) + \cdots + (a_i + a_1) \geqslant i a_{i+1}.$$

So

$$s_i \geqslant \frac{i}{2} \cdot a_{i+1}.$$

Thus,

$$\begin{aligned}\sum_{i=1}^{n}\frac{a_i}{i} &= \sum_{i=1}^{n}\frac{s_i - s_{i-1}}{i}\\ &= \sum_{i=1}^{n-1}s_i\left(\frac{1}{i}-\frac{1}{i+1}\right)+\frac{1}{n}\cdot s_n\\ &\geqslant \frac{1}{2}s_1+\sum_{i=1}^{n-1}\frac{ia_{i+1}}{2}\left(\frac{1}{i}-\frac{1}{i+1}\right)+\frac{1}{n}s_n\\ &= \frac{1}{2}s_1+\frac{1}{2}\sum_{i=1}^{n-1}\frac{a_{i+1}}{i+1}+\frac{1}{n}s_n\\ &= \frac{1}{2}\sum_{i=1}^{n}\frac{a_i}{i}+\frac{1}{n}s_n.\end{aligned}$$

Hence

$$\begin{aligned}\sum_{i=1}^{n}\frac{a_i}{i} \geqslant \frac{2}{n}s_n &= \frac{2}{n}(s_{n-1}+a_n)\\ &\geqslant \frac{2}{n}\left(\frac{n-1}{2}a_n+a_n\right)\\ &= \frac{n+1}{n}a_n > a_n.\end{aligned}$$

Therefore, the original inequality holds.

Note. If we ignore the fact that a_1, a_2, $\cdots$ are real, we can prove this problem using mathematical induction. (See exercise 7, problem 5)

Eg. 9 (Rearrangement Inequality) Assume we have two ordered sets: $a_1 \leqslant a_2 \leqslant \cdots \leqslant a_n$ and $b_1 \leqslant b_2 \leqslant \cdots \leqslant b_n$. Prove that

$$\begin{aligned}& a_1b_1 + a_2b_2 + \cdots + a_nb_n \text{ (similarly sorted)}\\ \geqslant\ & a_1b_{j_1} + a_2b_{j_2} + \cdots + a_nb_{j_n} \text{ (randomly sorted)}\\ \geqslant\ & a_1b_n + a_2b_{n-1} + \cdots + a_nb_1 \text{ (oppositely sorted)},\end{aligned}$$

where j_1, j_2, $\cdots$, j_n is a permutation of 1, 2, $\cdots$, n.

Proof. Denote

$$s_i = b_1 + b_2 + \cdots + b_i,$$

$$s_i' = b_{j_1} + b_{j_2} + \cdots + b_{j_i}\ (i = 1, 2, \cdots, n).$$

From the conditions, we know $s_i \leqslant s'_i$, $(i = 1, 2, \cdots, n-1)$ and $s_n = s'_n$.

Since $a_i - a_{i+1} \leqslant 0$, we have

$$s_i(a_i - a_{i+1}) \geqslant s'_i(a_i - a_{i+1}).$$

Hence,

$$\begin{aligned}\sum_{i=1}^{n} a_i b_i &= \sum_{i=1}^{n-1} s_i(a_i - a_{i+1}) + a_n s_n \\ &\geqslant \sum_{i=1}^{n-1} s'_i(a_i - a_{i+1}) + a_n s'_n \\ &= \sum_{i=1}^{n} a_i b_{j_i}.\end{aligned}$$

Thus the left hand side of the original inequality is proved. Similarly, we can prove the right hand side of the inequality (we ask our readers to try themselves).

Eg. 10 Show that for every positive integer n,

$$\frac{2n+1}{3}\sqrt{n} \leqslant \sum_{i=1}^{n} \sqrt{i} \leqslant \frac{4n+3}{6}\sqrt{n} - \frac{1}{6}.$$

The equal signs on both sides hold if and only if $n = 1$.

Hint. We can treat $\sqrt{i}$ as either $1 \cdot \sqrt{i}$ or $i \cdot \frac{1}{\sqrt{i}}$, which will give us two estimations.

Proof. It is easy to verify that when $n = 1$, both sides of the inequality reaches equality. From now on we assume $n \geqslant 2$.

First let us prove the left hand side. Let $a_i = 1$, $b_i = \sqrt{i}$, then

$$s_i = a_1 + a_2 + \cdots + a_i = i.$$

Applying Abel's summation by parts formula, we obtain

$$\begin{aligned}s &= \sum_{i=1}^{n} \sqrt{i} = \sum_{i=1}^{n} a_i b_i \\ &= \sum_{i=1}^{n-1} i(\sqrt{i} - \sqrt{i+1}) + n\sqrt{n}\end{aligned}$$

$$= n\sqrt{n} - \sum_{i=1}^{n-1} \frac{i}{\sqrt{i} + \sqrt{i+1}}.$$

From $\frac{i}{\sqrt{i} + \sqrt{i+1}} < \frac{i}{2\sqrt{i}} = \frac{1}{2}\sqrt{i}$, we have

$$\sum_{i=1}^{n-1} \frac{i}{\sqrt{i} + \sqrt{i+1}} < \frac{1}{2}\sum_{i=1}^{n-1}\sqrt{i} = \frac{1}{2}(s - \sqrt{n}).$$

Thus

$$s > n\sqrt{n} - \frac{1}{2}(s - \sqrt{n}).$$

Solving s, we have

$$s > \frac{2n+1}{3} \cdot \sqrt{n}.$$

Next we prove the right hand side of the inequality.

Denote $a_i = i$, $b_i = \frac{1}{\sqrt{i}} (1 \leqslant i \leqslant n)$, then

$$s_i = a_1 + a_2 + \cdots + a_i = \frac{1}{2}i(i+1).$$

Again, using Abel's summation by parts formula, we have

$$\begin{aligned} s &= \sum_{i=1}^{n-1} \frac{i(i+1)}{2}\left(\frac{1}{\sqrt{i}} - \frac{1}{\sqrt{i+1}}\right) + \frac{1}{\sqrt{n}} \cdot \frac{n(n+1)}{2} \\ &= \frac{n+1}{2} \cdot \sqrt{n} + \sum_{i=1}^{n-1} \frac{i(i+1)}{2} \cdot \frac{1}{\sqrt{i(i+1)}(\sqrt{i} + \sqrt{i+1})} \\ &= \frac{n+1}{2} \cdot \sqrt{n} + \frac{1}{2}\sum_{i=1}^{n-1} \frac{\sqrt{i(i+1)}}{\sqrt{i} + \sqrt{i+1}}. \end{aligned}$$

Since

$$\frac{\sqrt{i(i+1)}}{\sqrt{i} + \sqrt{i+1}} < \frac{1}{4}(\sqrt{i} + \sqrt{i+1}) \ (i = 1, 2, \cdots, n-1).$$

We have

$$\sum_{i=1}^{n-1} \frac{\sqrt{i(i+1)}}{\sqrt{i}+\sqrt{i+1}} < \frac{1}{4} \cdot \sum_{i=1}^{n-1} (\sqrt{i}+\sqrt{i+1}) = \frac{1}{4}(2s-\sqrt{n}-1).$$

Thus

$$s < \frac{1}{8}(2s-\sqrt{n}-1) + \frac{n+1}{2} \cdot \sqrt{n}.$$

So

$$s < \frac{4n+3}{6} \cdot \sqrt{n} - \frac{1}{6}.$$

The right hand side is therefore proved.

Exercise 2

1. Let a, b, p, q be positive real numbers. Prove that

$$\frac{a^{p+q}+b^{p+q}}{2} \geqslant \left(\frac{a^p+b^p}{2}\right)\left(\frac{a^q+b^q}{2}\right).$$

2. Assume $0 < a_1 \leqslant a_2 \leqslant \cdots \leqslant a_n$; $0 < b_1 \leqslant b_2 \leqslant \cdots \leqslant b_n$, then

$$\frac{\sum_{i=1}^{n} a_i^2 b_i}{\sum_{i=1}^{n} a_i b_i} \geqslant \frac{\sum_{i=1}^{n} a_i^2}{\sum_{i=1}^{n} a_i}.$$

3. Let n be a given positive integer, $n \geqslant 3$. Given n positive real numbers a_1, a_2, $\cdots$, a_n. Denote m the minimum of $|a_i - a_j|$ $(1 \leqslant i < j \leqslant n)$. Find the maximum for m given that $a_1^2 + a_2^2 + \cdots + a_n^2 = 1$.

4. Find all the positive integers a_1, a_2, $\cdots$, a_n, such that

$$\frac{99}{100} = \frac{a_0}{a_1} + \frac{a_1}{a_2} + \cdots + \frac{a_{n-1}}{a_n},$$

where $a_0 = 1$, and $(a_{k+1} - 1)a_{k-1} \geqslant a_k^2 - 1$ $(k = 1, 2, \cdots, n-1)$.

5. Assume a_1, a_2, $\cdots$, a_n are n positive numbers not all equal, and they satisfy $\sum_{k=1}^{n} a_k^{-2n} = 1$. Show that

$$\sum_{k=1}^{n} a_k^{2n} - n^2 \sum_{1 \leqslant i < j \leqslant n} \left(\frac{a_i}{a_j} - \frac{a_j}{a_i}\right)^2 > n^2.$$

6. Assume $n (\geqslant 2)$ real numbers $a_1 \leqslant a_2 \leqslant \cdots \leqslant a_n$. Denote $x = \frac{1}{n}\sum_{i=1}^{n} a_i$, $y = \frac{1}{n}\sum_{i=1}^{n} a_i^2$. Prove that

$$2\sqrt{y - x^2} \leqslant a_n - a_1 \leqslant \sqrt{2n(y - x^2)}.$$

7. Assume the inequality

$$\begin{aligned} &| z'_1 - z'_2 | + | z'_2 - z'_3 | + \cdots + | z'_{n-1} - z'_n | \\ &\leqslant \lambda \cdot [\, | z_1 - z_2 | + | z_2 - z_3 | + \cdots + | z_{n-1} - z_n | \,] \end{aligned}$$

holds for all complex numbers $z_1, z_2, \cdots, z_n$ that are not all equal. Here $kz'_k = \sum_{j=1}^{k} z_j$, $k = 1, 2, \cdots, n$. Prove that $\lambda \geqslant 1 - \frac{1}{n}$. When does the equality hold?

8. Assume $n (\geqslant 3)$ is a positive integer, $x_1, x_2, \cdots, x_n$ are real numbers. For $1 \leqslant i \leqslant n-1$, we have $x_i < x_{i+1}$. Prove the following inequality:

$$\frac{n(n-1)}{2} \cdot \sum_{1 \leqslant i < j \leqslant n} x_i x_j > \left[\sum_{i=1}^{n-1} (n-i) x_i\right]\left[\sum_{j=2}^{n} (j-1) x_j\right].$$

9. Assume $a_1 \geqslant a_2 \geqslant \cdots \geqslant a_n \geqslant a_{n+1} = 0$ are real numbers. Prove that

$$\sqrt{\sum_{k=1}^{n} a_k} \leqslant \sum_{k=1}^{n} \sqrt{k}\,(\sqrt{a_k} - \sqrt{a_{k+1}}).$$

10. Prove the Chebyshev Inequality: Assume $a_1 \leqslant a_2 \leqslant \cdots \leqslant a_n$, $b_1 \leqslant b_2 \leqslant \cdots \leqslant b_n$, then

$$n \cdot \sum_{k=1}^{n} a_k b_k \geqslant \left(\sum_{k=1}^{n} a_k\right) \cdot \left(\sum_{k=1}^{n} b_k\right) \geqslant n \sum_{k=1}^{n} a_k b_{n-k+1}.$$

11. (A Generalization of the W. Janous Inequality) Assume a_1,

$a_2, \cdots, a_n \in \mathbf{R}^+, p, q \in \mathbf{R}^+$. Denote $S = (a_1^p + a_2^p + \cdots + a_n^p)^{\frac{1}{p}}$. Then, for any permutation $i_1, i_2, \cdots, i_n$ of $1, 2, \cdots, n$, we have

$$\sum_{k=1}^{n} \frac{a_k^q - a_{i_k}^q}{S^p - a_k^p} \geqslant 0.$$

12. Assume $a_1, a_2, \cdots$ is a real number series satisfying $\sum_{j=1}^{n} a_j \geqslant \sqrt{n}$ for every $n \geqslant 1$. Show that for every $n \geqslant 1$, we have

$$\sum_{j=1}^{n} a_j^2 \geqslant \frac{1}{4}\left(1 + \frac{1}{2} + \cdots + \frac{1}{n}\right).$$

13. Prove that for every real number x, the following inequality holds:

$$\sum_{k=1}^{n} \frac{[kx]}{k} \leqslant [nx],$$

where $[x]$ denotes the largest integer not exceeding x.

14. Assume $a_k \geqslant 0$, $k = 1, 2, \cdots, n$. Define $A_k = \frac{1}{k} \cdot \sum_{i=1}^{k} a_i$. Show that

$$\sum_{k=1}^{n} A_k^2 \leqslant 4 \sum_{k=1}^{n} a_k^2.$$

Chapter 3 Substitution Method

Substitution method is one of the most frequently used problem solving techniques. Regarding a complicated expression as a whole and replacing it with a symbol may dramatically simplify the problem. For instance, sometimes we can use trigonometric transformations: when we detect forms like "$x^2 + y^2 = r^2$", "$x^2 + y^2 \leqslant r^2$", "$\sqrt{r^2 - x^2}$" or "$|x| < 1$" in the problem, we can try to replace the variables using $\sin\alpha$ or $\cos\alpha$; when we detect forms like "$\sqrt{r^2 + x^2}$", "$\sqrt{x^2 - r^2}$", we can replace x by $r\tan\alpha$ or $r\sec\alpha$. During the substitution, we have to pay attention to the range of α, which is determined by the range of the variable being replaced.

Eg. 1 Assume $0° \leqslant \alpha \leqslant 90°$, prove that

$$2 \leqslant \sqrt{5 - 4\sin\alpha} + \sin\alpha \leqslant \frac{9}{4}.$$

Proof. Write $x = \sqrt{5 - 4\sin\alpha}$, then $\sin\alpha = \frac{5 - x^2}{4}$. Since $0 \leqslant \sin\alpha \leqslant 1$, we have $1 \leqslant x \leqslant \sqrt{5}$.

Denote

$$\begin{aligned} y &= \sqrt{5 - 4\sin\alpha} + \sin\alpha \\ &= x + \frac{5 - x^2}{4} \\ &= -\frac{1}{4}(x - 2)^2 + \frac{9}{4}. \end{aligned}$$

From $1 \leqslant x \leqslant \sqrt{5}$ we know that $2 \leqslant y \leqslant \frac{9}{4}$, therefore

$$2 \leqslant \sqrt{5-4\sin\alpha} + \sin\alpha \leqslant \frac{9}{4}.$$

Note. By replacing $x = \sqrt{5-4\sin\alpha}$, we can thus change the original inequality into a quadratic expression and simplify the problem.

Eg. 2 Assume that real numbers x and y satisfy $x^2+y^2-4x-6y+9=0$. Prove that

$$19 \leqslant x^2+y^2+12x+6y \leqslant 99.$$

Proof. Rewriting the condition, we have

$$(x-2)^2+(y-3)^2=4,$$

or

$$\left(\frac{x-2}{2}\right)^2+\left(\frac{y-3}{2}\right)^2=1.$$

Denote $\frac{x-2}{2}=\cos\theta$, $\frac{y-3}{2}=\sin\theta$, where $\theta\in[0, 2\pi)$, thus

$$\begin{aligned}
&x^2+y^2+12x+6y\\
&=(4x+6y-9)+12x+6y\\
&=16x+12y-9\\
&=16(2\cos\theta+2)+12(2\sin\theta+3)-9\\
&=32\cos\theta+24\sin\theta+59=8(4\cos\theta+3\sin\theta)+59\\
&=40\cos(\theta+\phi)+59,
\end{aligned}$$

where $\tan\phi=\frac{3}{4}$.

Since $19\leqslant 40\cos(\theta+\phi)+59\leqslant 99$, we obtain

$$19 \leqslant x^2+y^2+12x+6y \leqslant 99.$$

Eg. 3 Assume a, b, c are three sides of a triangle. Prove that

$$a^2b(a-b)+b^2c(b-c)+c^2a(c-a)\geqslant 0.$$

Proof. Denote $a=y+z$, $b=z+x$, $c=x+y$, $x, y, z\in\mathbf{R}^+$. Then, the original inequality is equivalent to

$$\begin{aligned}
&(y+z)^2(z+x)(y-x)+(z+x)^2(x+y)(z-y)\\
&\quad+(x+y)^2(y+z)(x-z)\geqslant 0.
\end{aligned}$$

$$\Leftrightarrow xy^3 + yz^3 + zx^3 \geqslant x^2yz + xy^2z + xyz^2.$$

$$\Leftrightarrow \frac{x^2}{y} + \frac{y^2}{z} + \frac{z^2}{x} \geqslant x + y + z.$$

Because

$$\frac{x^2}{y} + y \geqslant 2x, \frac{y^2}{z} + z \geqslant 2y, \frac{z^2}{x} + x \geqslant 2z,$$

we have

$$\frac{x^2}{y} + \frac{y^2}{z} + \frac{z^2}{x} \geqslant x + y + z.$$

Hence the original inequality holds true.

Note. When handling an inequality that deals with the three sides a, b, c of a triangle, we can make substitutions $a = y + z$, $b = z + x$, $c = x + y$, where x, y, $z \in \mathbf{R}^+$. In fact, as shown in Figure 3.1, the inscribed circle of $\triangle ABC$ is tangent to the sides BC, CA, AB at points D, E, F respectively. Let $AE = AF = x$, $BD = BF = y$, $CD = CE = z$, then $a = y + z$, $b = z + x$, $c = x + y$. With substitution, the inequality with respect to a, b, c is transformed to an inequality with respect to positive real numbers x, y, z.

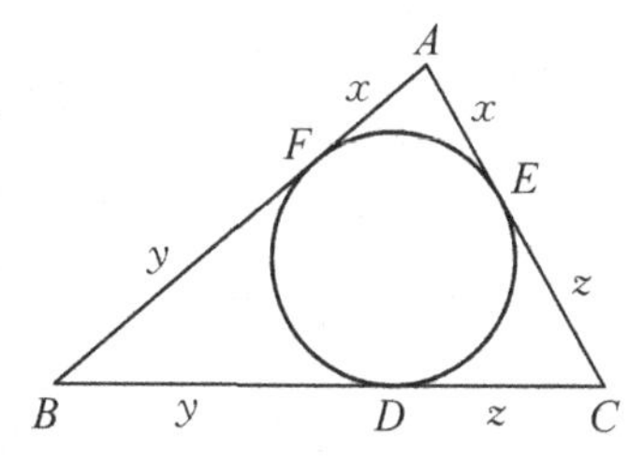

Figure 3.1

Eg. 4 Let a, b, c, d be positive real numbers such that

$$\frac{a^2}{1 + a^2} + \frac{b^2}{1 + b^2} + \frac{c^2}{1 + c^2} + \frac{d^2}{1 + d^2} = 1.$$

Prove that

$$abcd \leqslant \frac{1}{9}.$$

Proof. Let $a = \tan\alpha$, $b = \tan\beta$, $c = \tan\gamma$, $d = \tan\delta$. α, β, γ, $\delta \in (0, \pi/2)$. We have:

$$\sin^2\alpha + \sin^2\beta + \sin^2\gamma + \sin^2\delta = 1.$$

Thus,

$$3 \cdot \sqrt[3]{\sin^2\alpha \sin^2\beta \sin^2\gamma} \leqslant \sin^2\alpha + \sin^2\beta + \sin^2\gamma = \cos^2\delta,$$
$$3 \cdot \sqrt[3]{\sin^2\alpha \sin^2\beta \sin^2\delta} \leqslant \sin^2\alpha + \sin^2\beta + \sin^2\delta = \cos^2\gamma,$$
$$3 \cdot \sqrt[3]{\sin^2\alpha \sin^2\gamma \sin^2\delta} \leqslant \sin^2\alpha + \sin^2\gamma + \sin^2\delta = \cos^2\beta,$$
$$3 \cdot \sqrt[3]{\sin^2\beta \sin^2\gamma \sin^2\delta} \leqslant \sin^2\beta + \sin^2\gamma + \sin^2\delta = \cos^2\alpha.$$

Multiplying the above four inequality, we obtain:

$$\tan^2\alpha \tan^2\beta \tan^2\gamma \tan^2\delta \leqslant \frac{1}{81}.$$

Therefore,

$$abcd = \tan\alpha \tan\beta \tan\gamma \tan\delta \leqslant \frac{1}{9}.$$

Eg. 5 Let a, b, c be positive real numbers such that $abc = 1$. Prove that

(1) $\dfrac{1}{1+2a} + \dfrac{1}{1+2b} + \dfrac{1}{1+2c} \geqslant 1$;

(2) $\dfrac{1}{1+a+b} + \dfrac{1}{1+b+c} + \dfrac{1}{1+c+a} \leqslant 1$.

Proof. (1) [Method 1] Assume $a = \dfrac{x}{y}$, $b = \dfrac{y}{z}$, $c = \dfrac{z}{x}$, $x, y, z \in \mathbf{R}^+$. Then the original inequality is equivalent to

$$s = \frac{y}{y+2x} + \frac{z}{z+2y} + \frac{x}{x+2z} \geqslant 1.$$

Applying Cauchy Inequality,

$$s \cdot [y(y+2x) + z(z+2y) + x(x+2z)] \geqslant (x+y+z)^2.$$

Hence $s \geqslant 1$.

[Method 2] First, let us prove that

$$\frac{1}{1+2a} \geqslant \frac{a^{-\frac{2}{3}}}{a^{-\frac{2}{3}} + b^{-\frac{2}{3}} + c^{-\frac{2}{3}}}. \qquad ①$$

① is equivalent to

$$b^{-\frac{2}{3}}+c^{-\frac{2}{3}}\geqslant 2a^{\frac{1}{3}}.$$

Since

$$b^{-\frac{2}{3}}+c^{-\frac{2}{3}}\geqslant 2\cdot(bc)^{-\frac{1}{3}}=2a^{\frac{1}{3}}.$$

① is thus proved.

Similarly, we have

$$\frac{1}{1+2b}\geqslant\frac{b^{-\frac{2}{3}}}{a^{-\frac{2}{3}}+b^{-\frac{2}{3}}+c^{-\frac{2}{3}}},\qquad ②$$

$$\frac{1}{1+2c}\geqslant\frac{c^{-\frac{2}{3}}}{a^{-\frac{2}{3}}+b^{-\frac{2}{3}}+c^{-\frac{2}{3}}}.\qquad ③$$

Adding ①, ② and ③, the original inequality holds.

(2) Let $a=x^3$, $b=y^3$, $c=z^3$, x, y, $z\in\mathbf{R}^+$. Then, the condition is translated as $xyz=1$.

Since $x^3+y^3\geqslant x^2y+y^2x$, we have

$$\begin{aligned}\frac{1}{1+a+b}&=\frac{1}{1+x^3+y^3}\leqslant\frac{1}{1+x^2y+xy^2}\\&=\frac{1}{xyz+x^2y+y^2x}=\frac{1}{xy(x+y+z)}\\&=\frac{z}{x+y+z}.\end{aligned}$$

Similarly,

$$\frac{1}{1+b+c}\leqslant\frac{x}{x+y+z},$$

$$\frac{1}{1+c+a}\leqslant\frac{y}{x+y+z}.$$

Adding the above three inequality, the original inequality is thus proved.

Note. When three numbers has product 1, the above mentioned two substitution methods are both quite helpful.

Eg. 6 Let x, y, z be positive real numbers such that $\frac{1}{x}+\frac{1}{y}+\frac{1}{z}=1$, prove that

$$\sqrt{x+yz}+\sqrt{y+zx}+\sqrt{z+xy} \geqslant \sqrt{xyz}+\sqrt{x}+\sqrt{y}+\sqrt{z}.$$

Proof. Let $x=\frac{1}{\alpha}$, $y=\frac{1}{\beta}$, $z=\frac{1}{\gamma}$, then $\alpha+\beta+\gamma=1$.

The original inequality is equivalent to

$$\sum \sqrt{\frac{1}{\alpha}+\frac{1}{\beta\gamma}} \geqslant \sqrt{\frac{1}{\alpha\beta\gamma}}+\sum \sqrt{\frac{1}{\alpha}},$$

or

$$\sum \sqrt{\alpha+\beta\gamma} \geqslant \sum \sqrt{\alpha^2}+\sum \sqrt{\beta\gamma}.$$

This is equivalent to

$$\sum \sqrt{\alpha \cdot \sum \alpha+\beta\gamma} \geqslant \sum \sqrt{\alpha^2}+\sum \sqrt{\beta\gamma},$$

or

$$\sum \sqrt{(\alpha+\beta)(\alpha+\gamma)} \geqslant \sum (\sqrt{\alpha^2}+\sqrt{\beta\gamma}).$$

However, it is to show that

$$\sqrt{(\alpha+\beta)(\alpha+\gamma)} \geqslant \sqrt{\alpha^2}+\sqrt{\beta\gamma}.$$

As a result, the original inequality is proved.

Eg. 7 Let x, y, z be positive real numbers, prove that

$$\frac{y^2-x^2}{z+x}+\frac{z^2-y^2}{x+y}+\frac{x^2-z^2}{y+z} \geqslant 0. \quad \text{(W. Janous Inequality)}$$

Hint. On the left hand side of the inequality we see a sum in each denominator, which is hard to handle, thus we consider to replace them and simplify the denominators first.

Proof. Let $x+y=c$, $y+z=a$, $z+x=b$, the original inequality is equivalent to

$$\frac{c(a-b)}{b}+\frac{a(b-c)}{c}+\frac{b(c-a)}{a}\geqslant 0,$$

or

$$\frac{ac^2(a-b)+a^2b(b-c)+b^2c(c-a)}{abc}\geqslant 0.$$

So it suffices to show that

$$a^2c^2+a^2b^2+b^2c^2-abc^2-a^2bc-ab^2c\geqslant 0.$$

This is quite obvious.

Eg. 8 Let x_1, x_2, x_3 be positive real numbers. Show that

$$x_1x_2x_3\geqslant(x_2+x_3-x_1)(x_1+x_3-x_2)(x_1+x_2-x_3).$$

Proof. WLOG, let us assume that $x_1\geqslant x_2\geqslant x_3>0$.

Let $x_1=x_3+\delta_1$, $x_2=x_3+\delta_2$, then $\delta_1\geqslant\delta_2\geqslant 0$. Hence,

$$\begin{aligned}
&x_1x_2x_3-(x_2+x_3-x_1)(x_1+x_3-x_2)(x_1+x_2-x_3)\\
&=(x_3+\delta_1)(x_3+\delta_2)x_3-(x_3+\delta_2-\delta_1)(x_3+\delta_1-\delta_2)(x_3+\delta_1+\delta_2)\\
&=(x_3^2+\delta_1x_3+\delta_2x_3+\delta_1\delta_2)x_3-[x_3^2-(\delta_1-\delta_2)^2](x_3+\delta_1+\delta_2)\\
&=x_3\delta_1\delta_2+x_3^2(x_3+\delta_1+\delta_2)-[x_3^2-(\delta_1-\delta_2)^2](x_3+\delta_1+\delta_2)\\
&=x_3\delta_1\delta_2+(\delta_1-\delta_2)^2(x_3+\delta_1+\delta_2)\\
&\geqslant 0.
\end{aligned}$$

Therefore,

$$x_1x_2x_3\geqslant(x_2+x_3-x_1)(x_1+x_3-x_2)(x_1+x_2-x_3).$$

Note. The substitution method we used here is called substitution by increments.

Eg. 9 Find the maximum positive integer n, such that there exist n different real numbers x_1, x_2, $\cdots$, x_n satisfying: for any $1\leqslant i<j\leqslant n$, we have

$$(1+x_ix_j)^2\leqslant 0.9(1+x_i)^2(1+x_j)^2.$$

Soln.

$$(1+x_ix_j)^2\leqslant 0.9(1+x_i)^2(1+x_j)^2$$

is equivalent to

$$0.1(x_i x_j + 1)^2 \leqslant 0.9(x_i - x_j)^2,$$

or

$$| x_i x_j + 1 | \leqslant 3 | x_i - x_j |.$$

Let $x_i = \tan \alpha_i (1 \leqslant i \leqslant n)$. WLOG assume that $0 \leqslant \alpha_1 < \cdots < \alpha_n < \pi$.

The original inequality is equivalent to

$$| \tan(\alpha_i - \alpha_j) | \geqslant \frac{1}{3},$$

or

$$\pi - \theta \geqslant \alpha_j - \alpha_i \geqslant \theta,$$

where $\theta = \arctan \frac{1}{3}$.

Hence, we need to find the maximum n, such that there exist n angles $0 \leqslant \alpha_1 \leqslant \alpha_2 < \cdots < \alpha_n < \pi$ satisfying:

$$\alpha_n - \alpha_1 \leqslant \pi - \theta, \text{ and } \alpha_{i+1} - \alpha_i \geqslant \theta.$$

Consider the complex number $z = 3 + \mathrm{i}$, then $\theta = \arg z$.

Since $z^8 = 16(-527 + 336\mathrm{i})$, $z^9 = 16(-1917 + 481\mathrm{i})$, $z^{10} = 16(-(1917 \times 3 + 481) - 474\mathrm{i})$, we know that the main argument of z^9 is $< \pi$ while the main argument of z^{10} is $> \pi$. Thus,

$$9\theta < \pi < 10\theta.$$

From $\pi - \theta \geqslant \alpha_n - \alpha_1 \geqslant (n-1)\theta$, we have $n \leqslant 9$.

Finally, when $\alpha_1 = 0$, $\alpha_2 = \theta$, $\alpha_3 = 2\theta$, $\cdots$, $\alpha_9 = 8\theta$, the equal sign can be reached. Therefore the maximum of n is 9.

Eg. 10 Let x, y, z be positive real numbers. Prove that

$$x(y + z - x)^2 + y(z + x - y)^2 + z(x + y - z)^2$$
$$\geqslant 2xyz\left(\frac{x}{y+z} + \frac{y}{z+x} + \frac{z}{x+y}\right),$$

the equality holds if and only if $x = y = z$.

Proof. Let $a = y + z - x$, $b = x + z - y$, $c = x + y - z$, then

$$x = \frac{b + c}{2},\ y = \frac{c + a}{2},\ z = \frac{a + b}{2}.$$

The original inequality is thus equivalent to

$$\frac{1}{2}\sum a^2(b + c) \geqslant 2\frac{(a + b)(b + c)(c + a)}{8} \cdot \sum \frac{\frac{b + c}{2}}{a + \frac{b + c}{2}},$$

or

$$\begin{aligned} &2[a^2(b + c) + b^2(c + a) + c^2(a + b)] \\ &\geqslant (a + b)(b + c)(c + a) \cdot \sum \frac{b + c}{2a + b + c}. \end{aligned}$$

Next, let us prove that

$$a^2(b + c) + b^2(c + a) \geqslant (a + b)(b + c)(c + a) \cdot \frac{a + b}{2c + a + b}. \quad ①$$

Note that ① is equivalent to

$$\frac{a^2}{a + c} + \frac{b^2}{b + c} \geqslant \frac{(a + b)^2}{a + b + 2c}.$$

Using Cauchy-Schwarz Inequality, this is quite obvious.

By symmetry, we have two other inequalities similar to ①. Adding them together, the original inequality is proved.

Eg. 11 Let a, b, c be positive real numbers, prove that

$$\frac{b^3}{a^2 + 8bc} + \frac{c^3}{b^2 + 8ca} + \frac{a^3}{c^2 + 8ab} \geqslant \frac{a + b + c}{9}.$$

Proof. Denote M the left hand side of the inequality. Let

$$\begin{aligned} S &= (a^2 + 8bc) + (b^2 + 8ca) + (c^2 + 8ab) \\ &= (a + b + c)^2 + 6(ab + bc + ca) \\ &\leqslant 3(a + b + c)^2. \end{aligned}$$

Thus,

$$3 = \frac{1}{M} \cdot \left(\frac{b^3}{a^2 + 8bc} + \frac{c^3}{b^2 + 8ca} + \frac{a^3}{c^2 + 8ab}\right) + \frac{1}{S}[(a^2 + 8bc)$$

$$+(b^2+8ca)+(c^2+8ab)]+\frac{1+1+1}{3}$$

$$=\sum\left[\frac{b^3}{M(a^2+8bc)}+\frac{a^2+8bc}{S}+\frac{1}{3}\right]$$

$$\geqslant 3\sqrt[3]{\frac{b^3}{3SM}}+3\sqrt[3]{\frac{a^3}{3SM}}+3\sqrt[3]{\frac{c^3}{3SM}}$$

$$=\frac{3(a+b+c)}{\sqrt[3]{3SM}}.$$

Hence $3SM \geqslant (a+b+c)^3$, therefore $M \geqslant \frac{1}{9}(a+b+c)$.

Eg. 12 Let a, b, c be non-negative real numbers with sum 1. Prove that

$$2 \leqslant (1-a^2)^2+(1-b^2)^2+(1-c^2)^2 \leqslant (1+a)(1+b)(1+c).$$

Point out when does each of the equal signs hold.

Proof. Assume $ab+bc+ca=m$, $abc=n$, then

$$(x-a)(x-b)(x-c)=x^3-x^2+mx-n.$$

Let $x=a$, we have $a^3=a^2-ma+n$. (Note that we have reduced order!) As a result,

$$\sum a^3=\sum a^2-m\sum a+3n.$$

$$\begin{aligned}\sum a^4&=\sum a^3-m\sum a^2+n\sum a\\&=(1-m)\sum a^2-m\sum a+n\sum a+3n\\&=(1-m)(1-2m)-m+n+3n.\end{aligned}$$

Thus,

$$\begin{aligned}\sum(1-a^2)^2&=3-2\sum a^2+\sum a^4\\&=3-2(1-2m)+2m^2-3m+1-m+4n\\&=2m^2+4n+2\geqslant 2.\end{aligned}$$

The equality holds when there are 2 zeros in a, b, c.

On the other hand,

$$(1+a)(1+b)(1+c)=2+m+n.$$

Then $2m^2 + 4n + 2 \leqslant 2 + m + n \Leftrightarrow 3n \leqslant m - 2m^2$, or

$$3abc \leqslant m(1 - 2m) = (ab + bc + ca)(a^2 + b^2 + c^2),$$

which is equivalent to

$$3 \leqslant \left(\frac{1}{a} + \frac{1}{b} + \frac{1}{c}\right)(a^2 + b^2 + c^2). \quad ①$$

According to the Cauchy-Schwarz Inequality,

$$3(a^2 + b^2 + c^2) \geqslant (a + b + c)^2 = a + b + c.$$

Hence

$$\begin{aligned} & 3(a^2 + b^2 + c^2)\left(\frac{1}{a} + \frac{1}{b} + \frac{1}{c}\right) \\ \geqslant & (a + b + c)\left(\frac{1}{a} + \frac{1}{b} + \frac{1}{c}\right) \\ \geqslant & 9. \end{aligned}$$

Therefore ① holds and the equal sign is reached when $a = b = c = \frac{1}{3}$.

Eg. 13 Let $n \in \mathbf{N}_+$, $x_0 = 0$, $x_i > 0 (i = 1, 2, \cdots, n)$, and $\sum_{i=1}^{n} x_i = 1$. Prove that

$$1 \leqslant \sum_{i=1}^{n} \frac{x_i}{\sqrt{1 + x_0 + \cdots + x_{i-1}}\sqrt{x_i + \cdots + x_n}} < \frac{\pi}{2}.$$

Proof. Since

$$\begin{aligned} & \sqrt{1 + x_0 + \cdots + x_{i-1}}\sqrt{x_i + \cdots + x_n} \\ \leqslant & \frac{1}{2}[(1 + x_0 + \cdots + x_{i-1}) + (x_i + \cdots + x_n)] = 1. \end{aligned}$$

Thus,

$$s_i = \frac{x_i}{\sqrt{1 + x_0 + \cdots + x_{i-1}}\sqrt{x_i + \cdots + x_n}} \geqslant x_i (1 \leqslant i \leqslant n).$$

So

$$s=\sum_{i=1}^{n}s_i \geqslant \sum_{i=1}^{n}x_i=1.$$

The left hand side of the inequality is proved.

On the other hand, since $0 \leqslant x_0+x_1+\cdots+x_i \leqslant 1$, $i=0, 1, 2, \cdots, n$, we denote

$$\theta_i=\arcsin(x_0+x_1+\cdots+x_i) \in \left[0, \frac{\pi}{2}\right], \ i=0, 1, \cdots, n,$$

$$0=\theta_0<\theta_1<\theta_2<\cdots<\theta_n=\frac{\pi}{2}.$$

Also,

$$\sin\theta_i=x_0+x_1+\cdots+x_i;$$

$$\sin\theta_{i-1}=x_0+x_1+\cdots+x_{i-1}.$$

Hence

$$\begin{aligned} x_i &= \sin\theta_i-\sin\theta_{i-1} \\ &= 2\cos\frac{\theta_i+\theta_{i-1}}{2}\sin\frac{\theta_i-\theta_{i-1}}{2}, \ i=1, 2, \cdots, n. \end{aligned}$$

Because $\cos\frac{\theta_i+\theta_{i-1}}{2}<\cos\frac{2\theta_{i-1}}{2}=\cos\theta_{i-1}$, also it is easy to see that when $\theta \in \left(0, \frac{\pi}{2}\right]$,

$$\tan\theta>\theta>\sin\theta.$$

So

$$x_i<2\cos\theta_{i-1}\cdot\left(\frac{\theta_i-\theta_{i-1}}{2}\right)=\cos\theta_{i-1}(\theta_i-\theta_{i-1}), \ 1 \leqslant i \leqslant n,$$

thus

$$\frac{x_i}{\cos\theta_{i-1}}<\theta_i-\theta_{i-1},$$

then

$$\sum_{i=1}^{n}\frac{x_i}{\cos\theta_{i-1}}<\sum_{i=1}^{n}(\theta_i-\theta_{i-1})=\theta_n-\theta_0=\frac{\pi}{2}.$$

However,

$$\cos\theta_{i-1} = \sqrt{1-\sin^2\theta_{i-1}} = \sqrt{1-(x_0+x_1+\cdots+x_{i-1})^2}$$
$$= \sqrt{1+x_0+\cdots+x_{i-1}}\,\sqrt{x_i+\cdots+x_n},$$

therefore $s < \frac{\pi}{2}$.

Exercise 3

1. Let a, b, c be positive real numbers, find the minimum for $\frac{a}{b+3c}+\frac{b}{8c+4a}+\frac{9c}{3a+2b}$.

2. If the eclipse $\frac{x^2}{m^2}+\frac{y^2}{n^2}=1$ $(m, n>0)$ passes through a point $p(a, b)$ $(ab\neq 0, |a|\neq|b|)$. Find the minimum for $m+n$.

3. Assume a, b, c are the three sides of a triangle. Prove that

$$\frac{c-a}{b+c-a}+\frac{a-b}{c+a-b}+\frac{b-c}{a+b-c}\leqslant 0.$$

4. Let x, y, z be positive real numbers, show that

$$\frac{x+y+z}{3}\cdot\sqrt[3]{xyz}\leqslant\left(\frac{x+y}{2}\cdot\frac{y+z}{2}\cdot\frac{z+x}{2}\right)^{\frac{2}{3}}.$$

5. Let x, y be positive real numbers, show that

$$\left(\frac{2x+y}{3}\cdot\frac{x+2y}{3}\right)^2\geqslant\sqrt{xy}\cdot\left(\frac{x+y}{2}\right)^3.$$

6. Assume real numbers a, b satisfy $ab>0$. Prove that

$$\sqrt[3]{\frac{a^2b^2(a+b)^2}{4}}\leqslant\frac{a^2+10ab+b^2}{12}.$$

Determine when the equality holds. More generally, for any real numbers a, b, show that $\sqrt[3]{\frac{a^2b^2(a+b)^2}{4}}\leqslant\frac{a^2+ab+b^2}{3}$.

7. Let a, b, c be positive real numbers with product 1. Prove that

$$\frac{1}{1+a+b}+\frac{1}{1+b+c}+\frac{1}{1+c+a}\leqslant\frac{1}{2+a}+\frac{1}{2+b}+\frac{1}{2+c}.$$

8. Let a, b, c, d, e be positive real numbers and $abcde=1$. Prove that

$$\frac{a+abc}{1+ab+abcd}+\frac{b+bcd}{1+bc+bcde}+\frac{c+cde}{1+cd+cdea}+\frac{d+dea}{1+de+deab}+\frac{e+eab}{1+ea+eabc}\geqslant\frac{10}{3}.$$

9. Let a, b, c be positive real numbers. Prove that

$$a^2+b^2+c^2\geqslant\frac{c(a^2+b^2)}{a+b}+\frac{b(c^2+a^2)}{c+a}+\frac{a(b^2+c^2)}{b+c}.$$

10. Let x, y, z be positive real numbers satisfy $xyz+x+z=y$, find the maximum for $p=\frac{2}{x^2+1}-\frac{2}{y^2+1}+\frac{3}{z^2+1}$.

11. Show that one can find 4 different pairs of positive numbers (a, b) $(a\neq b)$, such that

$$\sqrt{(1-a^2)(1-b^2)}>\frac{a}{2b}+\frac{b}{2a}-ab-\frac{1}{8ab}.$$

12. Denote s the set of all triangles satisfying the following condition:

$$5\left(\frac{1}{AP}+\frac{1}{BQ}+\frac{1}{CR}\right)-\frac{3}{\min\{AP, BQ, CR\}}=\frac{6}{r}.$$

Here r is the radius of the inscribe triangle of $\triangle ABC$. The inscribed triangle is tangent to sides AB, BC, CA at points P, Q, R respectively. Show that all the element in set s is isosceles triangle and that they are similar to each other.

13. Let a, b, c be positive real numbers and $abc=1$. Prove that

$$\left(a-1+\frac{1}{b}\right)\left(b-1+\frac{1}{c}\right)\left(c-1+\frac{1}{a}\right)\leqslant 1.$$

14. Let a, b, c be positive real numbers, prove that

$$\frac{a}{\sqrt{a^2+8bc}}+\frac{b}{\sqrt{b^2+8ac}}+\frac{c}{\sqrt{c^2+8ab}}\geqslant 1.$$

Chapter 4 Proof by Contradiction

Proof by contradiction is a strong method of proving statements. Some problems are hard to prove starting from the given assumptions, in which cases we should try using proof by contradiction because by assuming what we want to prove is false, we add in one or even more conditions that might help us in solving the problem. In this chapter, we focus on the application of proof by contradiction in solving inequalities.

Eg. 1 Prove that for any real numbers x, y, z, the following three inequalities cannot hold true at the same time:

$$|x|<|y-z|,\ |y|<|z-x|,\ |z|<|x-y|.$$

Proof. Assume by contradiction, say the three inequalities all hold. Thus,

$$x^2<(y-z)^2,$$
$$y^2<(z-x)^2,$$
$$z^2<(x-y)^2.$$

So

$$(x-y+z)(x+y-z)<0,$$
$$(y-z+x)(y+z-x)<0,$$
$$(z-x+y)(z+x-y)<0.$$

Multiplying the previous three inequalities, we have

$$(x+y-z)^2(y+z-x)^2(z+x-y)^2<0,$$

which leads to contradiction.

Eg. 2 Let a, b, c, d be non-negative integers such that $(a+b)^2+$

$3a+2b=(c+d)^2+3c+2d$. Prove that

$$a=c,\ b=d.$$

Proof. First we prove that $a+b=c+d$ by assuming the contradiction.

If $a+b\neq c+d$, WLOG we assume $a+b>c+d$, then $a+b\geqslant c+d+1$ since they are all integers. Moreover,

$$\begin{aligned}(a+b)^2+3a+2b&=(a+b)^2+2(a+b)+a\\&\geqslant(c+d+1)^2+2(c+d+1)+a\\&=(c+d)^2+4(c+d)+3+a\\&>(c+d)^2+3c+2d.\end{aligned}$$

This is a contradiction. Therefore $a+b=c+d$. Plug it in the original equation we obtain $a=c$, then $b=d$ also follows.

Hint. For integers x, y, if $x>y$, we have $x\geqslant y+1$. This property is useful in dealing with integer related inequalities.

Eg. 3 Assume there are 12 real numbers a_1, a_2, $\cdots$, a_{12} satisfying:

$$\begin{gathered}a_2(a_1-a_2+a_3)<0,\\a_3(a_2-a_3+a_4)<0,\\\cdots\cdots\\a_{11}(a_{10}-a_{11}+a_{12})<0.\end{gathered}$$

Show that one can find at least three positive numbers and three negative numbers in these 12 reals.

Proof. Assume by contradiction that there are at most 2 negative numbers in a_1, a_2, $\cdots$, a_{12}. Then there exists $1\leqslant k\leqslant 9$ such that a_k, a_{k+1}, a_{k+2}, a_{k+3} are all non-negatives.

According to the conditions,

$$\begin{gathered}a_{k+1}(a_k-a_{k+1}+a_{k+2})<0,\\a_{k+2}(a_{k+1}-a_{k+2}+a_{k+3})<0.\end{gathered}$$

Since $a_{k+1}\geqslant0$, $a_{k+2}\geqslant0$, we know $a_{k+1}>0$, $a_{k+2}>0$, and

$$a_k - a_{k+1} + a_{k+2} < 0,$$
$$a_{k+1} - a_{k+2} + a_{k+3} < 0.$$

Adding the previous two inequalities together, we obtain $a_k + a_{k+3} < 0$, which is a contradiction to the fact that $a_k \geqslant 0$ and $a_{k+3} \geqslant 0$. Thus, there are at least three negatives in $a_1, a_2, \cdots, a_{12}$. Similarly, we can prove that there are also at least three positives in $a_1, a_2, \cdots, a_{12}$.

Eg. 4 Assume that positive integers a, b, c, d, n satisfy

$$n^2 < a < b < c < d < (n+1)^2.$$

Prove that $ad \neq bc$.

Proof. Assume by contradiction that $\frac{b}{a} = \frac{d}{c}$.

Denote $\frac{b}{a} = \frac{d}{c} = \frac{p}{q}$, where p, q are two coprime positive integers.

Since $\frac{p}{q} > 1$, we have $p \geqslant q + 1$, hence

$$\frac{p}{q} \geqslant 1 + \frac{1}{q}. \quad ①$$

Also, as $b = \frac{ap}{q}$, $q \mid ap$. Thus $q \mid a$. Similarly $q \mid c$. Then we have $q \mid c - a$, so $c - a \geqslant q$, $c \geqslant a + q$. As a result,

$$\frac{p}{q} = \frac{d}{c} \leqslant \frac{d}{a+q} < \frac{(n+1)^2}{n^2+q}. \quad ②$$

From ① and ② we know that

$$\frac{(n+1)^2}{n^2+q} > 1 + \frac{1}{q},$$

or

$$2n > q + \frac{n^2}{q}.$$

This is a contradiction! Therefore $ad \neq bc$.

Eg. 5 Assume a, b, c are positive real numbers that satisfy $a+b+c\geqslant abc$. Prove that at least two of the following inequalities hold true:

$$\frac{6}{a}+\frac{3}{b}+\frac{2}{c}\geqslant 2,\ \frac{6}{b}+\frac{3}{c}+\frac{2}{a}\geqslant 2,\ \frac{6}{c}+\frac{3}{a}+\frac{2}{b}\geqslant 2.$$

Proof. We assume the contradictory of the conclusion. We have two possibilities.

(i) If

$$\frac{6}{a}+\frac{3}{b}+\frac{2}{c}<2,\ \frac{6}{b}+\frac{3}{c}+\frac{2}{a}<2,\ \frac{6}{c}+\frac{3}{a}+\frac{2}{b}<2,$$

then

$$11\left(\frac{1}{a}+\frac{1}{b}+\frac{1}{c}\right)<6,$$

which is a contradiction since

$$\left(\frac{1}{a}+\frac{1}{b}+\frac{1}{c}\right)^2\geqslant 3\left(\frac{1}{ab}+\frac{1}{bc}+\frac{1}{ca}\right)\geqslant 3.$$

(ii) Now we consider the other possibility. WLOG assume two of the listed inequalities are false:

$$\frac{6}{a}+\frac{3}{b}+\frac{2}{c}<2, \quad ①$$

$$\frac{6}{b}+\frac{3}{c}+\frac{2}{a}<2, \quad ②$$

$$\frac{6}{c}+\frac{3}{a}+\frac{2}{b}\geqslant 2. \quad ③$$

From ① $\times 1+$ ② $\times 7-$ ③ $\times 1$, we have

$$\frac{43}{c}+\frac{17}{b}+\frac{17}{a}<14.$$

However, the left hand side of the above inequality $>17\left(\frac{1}{a}+\frac{1}{b}+\frac{1}{c}\right)\geqslant 17\sqrt{3}>14$, this leads us to a contradiction!

Eg. 6 Let x, y, z be positive real numbers. Prove that

$$\sqrt{x+\sqrt[3]{y+\sqrt[4]{z}}} \geqslant \sqrt[32]{xyz}.$$

Proof. Assume the reverse is true: i. e. assume that there exist positive numbers x_0, y_0, z_0, such that $\sqrt{x_0+\sqrt[3]{y_0+\sqrt[4]{z_0}}} < \sqrt[32]{x_0y_0z_0}$, thus we have:

$$\sqrt{x_0} < \sqrt[32]{x_0y_0z_0},$$

$$\sqrt[6]{y_0} < \sqrt[32]{x_0y_0z_0},$$

$$\sqrt[24]{z_0} < \sqrt[32]{x_0y_0z_0}.$$

Or

$$x_0^{16} < x_0y_0z_0,$$

$$y_0^{16} < (x_0y_0z_0)^3,$$

$$z_0^{16} < (x_0y_0z_0)^{12}.$$

Multiplying the above three inequalities, we obtain $x_0^{16}y_0^{16}z_0^{16} < (x_0y_0z_0)^{16}$, which is clearly a contradiction. Hence the original inequality holds.

Eg. 7 Assume for any real number x we have $\cos(a\sin x) > \sin(b\cos x)$, prove that

$$a^2+b^2 < \frac{\pi^2}{4}.$$

Proof. Assume by contradiction that $a^2+b^2 \geqslant \frac{\pi^2}{4}$. We can write $a\sin x + b\cos x$ as $\sqrt{a^2+b^2}\sin(x+\phi)$, where $\cos\phi = \frac{a}{\sqrt{a^2+b^2}}$, $\sin\phi = \frac{b}{\sqrt{a^2+b^2}}$.

Since $\sqrt{a^2+b^2} \geqslant \frac{\pi}{2}$, there exists real number x_0, such that

$$\sqrt{a^2+b^2}\sin(x_0+\phi) = \frac{\pi}{2}.$$

Or

$$a\sin x_0 + b\cos x_0 = \frac{\pi}{2}.$$

From the above we have $\cos(a\sin x_0) = \sin(b\cos x_0)$, which is a contradiction.

Therefore, $a^2 + b^2 < \frac{\pi^2}{4}$.

Eg. 8 Place an integer on every rational point on the real axis. Show that we can find an interval such that the sum of the integers on the end points of this interval is not greater than twice the integer placed on the midpoint of the interval.

Proof. Assume by contradiction that there exist a placing of the integers such that for any interval $[A, B]$ that contains point C as the midpoint, we have $c < \frac{a+b}{2}$, where a, b, c represent the integers placed on A, B, C respectively.

We name the points -1, 1, 0, $-\frac{1}{2^n}$, and $\frac{1}{2^n}$ as A, B, C, A_n and $B_n (n = 1, 2, 3, \cdots)$. Assume the integers placed on them are a, b, c, a_n, b_n alternatively:

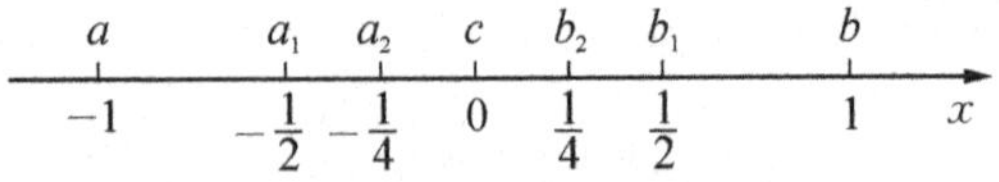

Hence, $a_1 < \frac{a+c}{2}$, $a_2 < \frac{a_1+c}{2}$.

So $\max\{a_1, a_2\} < \max\{a, c\}$.

Similarly, we have

$$\max\{a, c\} > \max\{a_1, a_2\} > \max\{a_3, a_4\} > \cdots,$$
$$\max\{b, c\} > \max\{b_1, b_2\} > \max\{b_3, b_4\} > \cdots.$$

Thus, there exists m, such that

$$a_{2m} \leqslant \max\{a, c\} - m, \; b_{2m} \leqslant \max\{b, c\} - m.$$

So

$$a_{2m}+b_{2m}\leqslant 1-2m<0.$$

However, 0 is the mid point of the interval $[a_{2m}, b_{2m}]$, this is a contradiction!

Eg. 9 Assume p is the product of two consecutive integers greater than 2. Prove that there exist no integers $x_1, x_2, \cdots, x_p$ satisfying the following equation:

$$\sum_{i=1}^{p}x_i^2-\frac{4}{4p+1}\left(\sum_{i=1}^{p}x_i\right)^2=1.$$

Proof. Write $p=k(k+1)$, $k\geqslant 3$. Then $p\geqslant 12$, $4p+1\geqslant 4p$. Assume by contradiction that there exist integers $x_1\geqslant x_2\geqslant\cdots\geqslant x_p$ satisfying the equation:

$$\begin{aligned}4p+1&=(4p+1)\sum_{i=1}^{p}x_i^2-4\left(\sum_{i}^{p}x_i\right)^2\\&=4\left[p\sum_{i=1}^{p}x_i^2-\left(\sum_{i}^{p}x_i\right)^2\right]+\sum_{i=1}^{p}x_i^2\\&=4\sum_{1\leqslant i<j\leqslant p}(x_i-x_j)^2+\sum_{i=1}^{p}x_i^2.\end{aligned}$$

If all x_i's are equal to each other, we have from above that

$$4p+1=px_1^2,$$

which is a contradiction. Hence we must have $x_1\geqslant x_2\geqslant\cdots\geqslant x_l>x_{l+1}\geqslant\cdots\geqslant x_p$, where $l\in\mathbf{N}_+$. Next we discuss two possibilities:

(i) If $2\leqslant l<p-1$,

$$\begin{aligned}4\sum_{1\leqslant i<j\leqslant p}(x_i-x_j)^2&\geqslant 4\sum_{i=1}^{l}\sum_{k=l+1}^{p}(x_i-x_k)^2\\&\geqslant 4l(p-l).\end{aligned}$$

Also,

$$\begin{aligned}l(p-l)-2(p-2)&=lp-l^2-2p+4\\&=(l-2)(p-l-2)\geqslant 0.\end{aligned}$$

We have:

$$4\sum_{1\leqslant i<j\leqslant p}(x_i-x_j)^2\geqslant 8(p-2)>4p+1,$$

which leads to a contradiction, so we consider the other possibility:

(ii) $l=1$ or $l=p-1$. WLOG we assume $l=1$, this implies

$$x_1>x_2=x_3=\cdots=x_{p-1}\geqslant x_p.$$

Thus,

$$\begin{aligned}4p+1&=4\sum_{s=2}^{p}(x_1-x_s)^2+4\sum_{s=2}^{p-1}(x_s-x_p)^2+\sum_{i=1}^{p}x_i^2\\&=4(p-2)(x_1-x_2)^2+4(x_1-x_p)^2\\&\quad+4(p-2)(x_2-x_p)^2+\sum_{i=1}^{p}x_i^2.\end{aligned}$$

Hence $x_1-x_2=1$, it follows that

$$9=4(x_1-x_p)^2+4(p-2)(x_2-x_p)^2+\sum_{i=1}^{p}x_i^2.$$

So $x_2=x_p(p\geqslant 12)$ and $x_1-x_p=1$. Therefore,

$$5=x_1^2+(p-1)^2x_2^2.$$

Since $p\geqslant 12$, we have $x_2=0$, then $x_1^2=5$, which is a contradiction.

Eg. 10 Let $a_1, a_2, \cdots, a_n$ be positive real numbers satisfying: $a_1+a_2+\cdots+a_n=\frac{1}{a_1}+\frac{1}{a_2}+\cdots+\frac{1}{a_n}$. Prove that

$$\frac{1}{n-1+a_1}+\frac{1}{n-1+a_2}+\cdots+\frac{1}{n-1+a_n}\geqslant 1.$$

Proof. Let $b_i=\frac{1}{n-1+a_i}$, $i=1, 2, \cdots, n$, then $b_i<\frac{1}{n-1}$, and

$$a_i=\frac{1-(n-1)b_i}{b_i},\ i=1,2,\cdots,n.$$

From the given condition, we have:

$$\sum_{i=1}^{n}\frac{1-(n-1)b_i}{b_i}=\sum_{i=1}^{n}\frac{b_i}{1-(n-1)b_i}.$$

Next assume by contradiction that

$$b_1 + b_2 + \cdots + b_n < 1. \qquad ①$$

From Cauchy-Schwarz Inequality, we obtain

$$\sum_{j \neq i}(1-(n-1)b_j) \cdot \sum_{j \neq i} \frac{1}{1-(n-1)b_j} \geqslant (n-1)^2.$$

According to ①,

$$\sum_{j \neq i}(1-(n-1)b_j) < (n-1)b_i,$$

so

$$\sum_{j \neq i} \frac{1}{1-(n-1)b_j} > \frac{n-1}{b_i}.$$

Hence,

$$\sum_{j \neq i} \frac{1-(n-1)b_i}{1-(n-1)b_j} > (n-1) \cdot \frac{1-(n-1)b_i}{b_i}.$$

Add up the above inequality for $i = 1, 2, \cdots, n$, we get

$$\sum_{i=1}^{n} \sum_{j \neq i} \frac{1-(n-1)b_i}{1-(n-1)b_j} > (n-1) \sum_{i=1}^{n} \frac{1-(n-1)b_i}{b_i},$$

or

$$\sum_{j=1}^{n} \sum_{j \neq i} \frac{1-(n-1)b_i}{1-(n-1)b_j} > (n-1) \sum_{i=1}^{n} \frac{1-(n-1)b_i}{b_i}. \qquad ②$$

However, due to ①,

$$\sum_{i \neq j}(1-(n-1)b_i) < b_j(n-1).$$

Thus by ②, we have

$$(n-1) \sum_{j=1}^{n} \frac{b_j}{1-(n-1)b_j} > (n-1) \sum_{i=1}^{n} \frac{1-(n-1)b_i}{b_i}.$$

This is a contradiction!

Eg. 11 For positive integer n $(n \geqslant 2)$, assume the coefficients of $f(x) = a_n x^n + a_{n-1}x^{n-1} + \cdots + a_1 x + a_0$ are all real numbers. Also, every complex root of $f(x)$ has a negative real part and $f(x)$ has a pair of equal real roots. Prove that there exists k, $1 \leqslant k \leqslant n-1$ such that

$$a_k^2 - 4a_{k-1}a_{k+1} \leqslant 0.$$

Proof. When $n=2$, $f(x) = a_2(x+a)^2$, where a is a positive real number. Hence,

$$f(x) = a_2(x^2 + 2ax + a^2).$$

So $a_1 = 2aa_2$, $a_0 = a^2 a_2$, $a_1^2 - 4a_0 a_2 = 0$, the conclusion is true.

Next we assume $n > 2$ and $a_n > 0$ (otherwise we multiply -1 to each of the coefficient). Then,

$$f(x) = (x+a)^2 (b_{n-2}x^{n-2} + b_{n-3}x^{n-3} + \cdots + b_1 x + b_0) \ (a \in \mathbf{R}^+). \quad ①$$

Since the complex roots of a real coefficient polynomial appear in pairs, we assume there are k conjugate complex roots, denoted them as $x_j \pm \mathrm{i}y_j$, $j = 1, 2, \cdots, k$. From the conditions we know $x_j < 0$ and the other $n-2-2k$ roots are all negative real roots, denote them as $z_1, z_2, \cdots, z_{n-2-2k}$. Therefore,

$$\begin{aligned}
&\frac{1}{b_{n-2}} \cdot f(x) \\
&= (x+a)^2 \cdot \prod_{j=1}^{k} [x - (x_j + iy_j)][x - (x_j - iy_j)] \cdot \prod_{l=1}^{n-2-2k} (x - z_l) \\
&= (x+a)^2 \cdot \prod_{j=1}^{k} (x^2 - 2x_j x + x_j^2 + y_j^2) \cdot \prod_{l=1}^{n-2-2k} (x - z_l). \quad ②
\end{aligned}$$

Combining ① and ②, we have $b_j > 0$, $j = 0, 1, 2, \cdots, n-2$, and

$$\begin{aligned}
f(x) &= (x+a)^2 (b_{n-2}x^{n-2} + b_{n-3}x^{n-3} + \cdots + b_1 x + b_0) \\
&= b_{n-2}x^n + (2ab_{n-2} + b_{n-3})x^{n-1} + (a^2 b_{n-2} \\
&\quad + 2ab_{n-3} + b_{n-4})x^{n-2} + \cdots + (a^2 b_2 + 2ab_1 \\
&\quad + b_0)x^2 + (a^2 b_1 + 2ab_0)x + a^2 b_0.
\end{aligned}$$

To get a more unified expression, we set $b_i = 0$ when $i < 0$ and $i > n-2$. It follows that:

$$a_j = a^2 b_j + 2ab_{j-1} + b_{j-2},\ j = 0, 1, 2, \cdots, n.$$

If $n = 3$, we can prove by assuming the contradiction:

Assume

$$a_1^2 - 4a_0 a_2 > 0,\ a_2^2 - 4a_1 a_3 > 0.$$

$$a_0 = a^2 b_0,\ a_1 = a^2 b_1 + 2ab_0.$$

Because $a_2 = 2ab_1 + b_0$, $a_3 = b_1$, we have

$$0 < a_1^2 - 4a_0 a_2 = (a^2 b_1 + 2ab_0)^2 - 4a^2 b_0 (2ab_1 + b_0)$$
$$= a^3 b_1 (ab_1 - 4b_0),$$

hence

$$ab_1 > 4b_0. \quad ①$$

On the other hand,

$$0 < a_2^2 - 4a_1 a_3 = (2ab_1 + b_0)^2 - 4(a^2 b_1 + 2ab_0) b_1$$
$$= b_0 (b_0 - 4ab_1),$$

so

$$b_0 > 4ab_1. \quad ②$$

① and ② are contradictions!

When $n > 3$, we also assume by contradiction:

Assume that for $k \in \{1, 2, \cdots, n-1\}$, we have $a_k^2 - 4a_{k-1} a_{k+1} > 0$, Then

$$0 < a_k^2 - 4a_{k-1} a_{k+1}$$
$$= (a^2 b_k + 2ab_{k-1} + b_{k-2})^2 - 4(a^2 b_{k-1} + 2ab_{k-2}$$
$$+ b_{k-3})(a^2 b_{k+1} + 2ab_k + b_{k-1})$$
$$< b_{k-2}^2 + a^4 b_k^2 - 4ab_{k-2} b_{k-1} - 4a^3 b_{k-1} b_k.$$

$$0 = a^3 b_k (ab_k - 4b_{k-1}) + b_{k-2}(b_{k-2} - 4ab_{k-1}).$$

Denote $q_k = ab_k - 4b_{k-1}$, $r_k = b_{k-1} - 4ab_k$, $k = 1, 2, \cdots, n-1$. Thus,

$$0 < a^3 b_k q_k + b_{k-2} r_{k-1}. \quad ③$$

If $k = 1$, $q_1 = ab_1 - 4b_0$.

From ③ we have $0 < a^3 b_1 q_1$, hence $q_1 > 0$. We have $ab_1 > 4b_0$, $r_1 = b_0 - 4ab_1 < 0$.

If $k = n - 1$, from ③ we have $b_{n-3} r_{n-2} > 0$, so $r_{n-2} > 0$.

Denote u the smallest subscript such that $r_u > 0 (2 \leqslant u \leqslant n - 2)$, so that $r_{u-1} \leqslant 0$.

In ③ we set $k = u$. Since $0 < a^3 b_u q_u + b_{u-2} r_{u-1}$, we have $q_u > 0$. Meanwhile,

$$q_u = ab_u - 4b_{u-1} > 0, \text{ so } ab_u > 4b_{u-1}. \quad ④$$

$$r_u = b_{u-1} - 4ab_u > 0, \text{ so } b_{u-1} > 4ab_u. \quad ⑤$$

Combining ④ and ⑤, we obtain a contradiction!

Exercise 4

1. Let a, b, c be real numbers such that $a + b + c > 0$, $ab + bc + ca > 0$, $abc > 0$. Prove that

$$a > 0, b > 0, c > 0.$$

2. Prove that the following inequality has no real solutions:

$$\begin{aligned} |x| &> |y - z + t|, \\ |y| &> |x - z + t|, \\ |z| &> |x - y + t|, \\ |t| &> |x - y + z|. \end{aligned}$$

3. Assume that real numbers a, b, c, d, p, q satisfy

$$ab + cd = 2pq, \ ac \geqslant p^2 > 0.$$

Prove that $bd \leqslant q^2$.

4. Assume that a, b, c are positive real numbers and $a + b + c \geqslant abc$. Prove that

$$a^2 + b^2 + c^2 \geqslant abc.$$

5. Assume that $f(x)$ and $g(x)$ are real functions based on $[0, 1]$. Prove that there exist x_0, $y_0 \in [0, 1]$, such that

$$|x_0 y_0 - f(x_0) - g(y_0)| \geqslant \frac{1}{4}.$$

6. Prove that for any real numbers a, b, there exists $x, y \in [0, 1]$, such that

$$|xy - ax - by| \geqslant \frac{1}{3}.$$

If we change the $\frac{1}{3}$ on the right hand side of the inequality into $\frac{1}{2}$ or $0.333\,34$, does the statement still hold true?

7. Let m, n be positive integers, a_1, a_2, $\cdots$, a_m are different elements in the set $\{1, 2, \cdots, n\}$.

Whenever $a_i + a_j \leqslant n$, $1 \leqslant i \leqslant j \leqslant m$, there exists some k, $1 \leqslant k \leqslant m$, such that $a_i + a_j = a_k$. Prove that

$$\frac{1}{m}(a_1 + a_2 + \cdots + a_m) \geqslant \frac{1}{2}(n+1).$$

8. For any $n \in \mathbf{Z}^+$ and a real sequence a_1, a_2, $\cdots$, a_n, prove that there exists $k \in \mathbf{Z}^+$, such that:

$$\left|\sum_{i=1}^{k} a_i - \sum_{i=k+1}^{n} a_i\right| \leqslant \max_{1 \leqslant i \leqslant n} |a_i|.$$

9. Assume x_k, $y_k \in \mathbf{R}$, $j_k = x_k + iy_k$ $(k = 1, 2, \cdots, n$, $i = \sqrt{-1})$. r is the absolute value of the real part of $\pm\sqrt{j_1^2 + j_2^2 + \cdots + j_n^2}$. Prove that

$$r \leqslant |x_1| + |x_2| + \cdots + |x_n|.$$

10. Given a non-increasing positive sequence $a_1 \geqslant a_2 \geqslant a_3 \geqslant \cdots \geqslant a_n \geqslant \cdots$, where $a_1 = \frac{1}{2k}$ $(k \in \mathbf{N}, k \geqslant 2)$ and $a_1 + a_2 + \cdots + a_n + \cdots = 1$. Show that we can pick k numbers in the sequence, such that the smallest one exceeds half of the largest one.

11. Prove or disprove the following statement: if x, y are real numbers and $y \geqslant 0$, $y(y+1) \leqslant (x+1)^2$, then $y(y-1) \leqslant x^2$.

12. Assume $a_1 \geqslant a_2 \geqslant \cdots \geqslant a_n$ are n real numbers satisfying the following condition: for any positive integer k, we have $a_1^k + a_2^k + \cdots + a_n^k \geqslant 0$. Let $p = \max\{|a_1|, |a_2|, \cdots, |a_n|\}$. Show that $p = a_1$, and for any $x > a_1$, we have

$$(x - a_1)(x - a_2)\cdots(x - a_n) \leqslant x^n - a_1^n.$$

13. Assume real numbers $a_1, a_2, \cdots, a_n (n \geqslant 2)$ and A satisfy $A + \sum_{i=1}^{n} a_i^2 < \frac{1}{n-1}\left(\sum_{i=1}^{n} a_i\right)^2$. Prove that for $1 \leqslant i < j \leqslant n$, we have $A < 2a_i a_j$.

14. Assume $\{a_k\}_{k=1}^{\infty}$ is an infinite sequence composed of non-negative real numbers that satisfy $a_k - 2a_{k+1} + a_{k+2} \geqslant 0$ and $\sum_{j=1}^{k} a_j \leqslant 1$, $k = 1, 2, \cdots$. Prove that

$$0 \leqslant a_k - a_{k+1} \leqslant \frac{2}{k^2}, \quad k = 1, 2, \cdots.$$

15. If the roots of the polynomial $x^4 + ax^3 + bx + c = 0$ are all real, prove that $ab \leqslant 0$.

Chapter 5 Construction Method

Construction method is an important approach in dealing with inequalities. By introducing a suitable identity, function, graph or series, we can turn the original statement into something more straightforward or fundamental, hence easier to prove.

5.1 Construct Identity

Identity can be viewed as the strongest inequality. Sometimes, if we can spot the omitted terms in the inequality, we might find the identity behind the inequality that is key to solving the problem.

Eg. 1 Let a, b, c, d be real numbers such that $a^2+b^2+c^2+d^2=1$. Prove that

$$(a+b)^4+(a+c)^4+(a+d)^4+(b+c)^4+(b+d)^4+(c+d)^4 \leqslant 6. \quad ①$$

Hint. From the assumption we know that $(a^2+b^2+c^2+d^2)^2=1$. We should try to find its relationship with the left hand side of ①. Note that in order to cancel odd orders of a in $(a+b)^4$ which does not appear in the expansion of $(a^2+b^2+c^2+d^2)^2$, we can add in terms like $(a-b)^4$.

Proof. Consider the sum

$$(a-b)^4+(a-c)^4+(a-d)^4+(b-c)^4+(b-d)^4+(c-d)^4.$$

It's easy to see that the above sum forms an identity with the left hand side of ①, i.e.

$$(a+b)^4+(a-b)^4+(a+c)^4+(a-c)^4+(a+d)^4+(a-d)^4$$

$$+(b+c)^4+(b-c)^4+(b+d)^4+(b-d)^4+(c+d)^4+(c-d)^4$$
$$=6(a^2+b^2+c^2+d^2)^2 \quad ②$$

② implies ① immediately.

Eg. 2 Assume $\triangle A_1A_2A_3$ and $\triangle B_1B_2B_3$ have sides of length a_1, a_2, a_3 and b_1, b_2, b_3 respectively and they have areas S_1, S_2 respectively. Denote

$$H=a_1^2(-b_1^2+b_2^2+b_3^2)+a_2^2(b_1^2-b_2^2+b_3^2)+a_3^2(b_1^2+b_2^2-b_3^2).$$

Then for any $\lambda \in \left\{\frac{b_1^2}{a_1^2}, \frac{b_2^2}{a_2^2}, \frac{b_3^2}{a_3^2}\right\}$, we have

$$H \geqslant 8\left(\lambda S_1^2+\frac{1}{\lambda}S_2^2\right).$$

Proof. According to Heron's Formula,

$$16S_1^2=2a_1^2a_2^2+2a_2^2a_3^2+2a_3^2a_1^2-a_1^4-a_2^4-a_3^4,$$

$$16S_2^2=2b_1^2b_2^2+2b_2^2b_3^2+2b_3^2b_1^2-b_1^4-b_2^4-b_3^4.$$

Set $D_1=\sqrt{\lambda}a_1^2-\sqrt{\frac{1}{\lambda}}b_1^2$, $D_2=\sqrt{\lambda}a_2^2-\sqrt{\frac{1}{\lambda}}b_2^2$, $D_3=\sqrt{\lambda}a_3^2-\sqrt{\frac{1}{\lambda}}b_3^2$. Thus, we have the identity:

$$H-8\left(\lambda S_1^2+\frac{1}{\lambda}S_2^2\right)$$
$$=\frac{1}{2}(D_1^2+D_2^2+D_3^2)-(D_1D_2+D_2D_3+D_3D_1). \quad ①$$

When $\lambda=\frac{b_1^2}{a_1^2}$, $D_1=0$. ① turns into

$$H-8\left(\lambda S_1^2+\frac{1}{\lambda}S_2^2\right)=\frac{1}{2}(D_2-D_3)^2.$$

So the conclusion is true.

Similarly, the inequality holds for $\lambda=\frac{b_2^2}{a_2^2}$ or $\frac{b_3^2}{a_3^2}$.

5.2 Construct Function

We can construct a proper function that caters to certain properties of the algebraic expression in an inequality. Utilizing the properties of linear functions and quadratic functions as well as the monotonicity of the functions will help us to prove the inequality.

Eg. 3 Assume $a, b, c \in (-2, 1)$, prove that

$$abc > a + b + c - 2.$$

Hint. The two sides of the inequality is symmetric with respect to a, b, c. Also, a, b, c are all of order 1, so we can try to construct a linear function.

Proof. Assume $f(x) = (bc - 1)x - b - c + 2$, we have

$$\begin{aligned} f(-2) &= -2bc - b - c + 4 \\ &= -2\left(b + \frac{1}{2}\right)\left(c + \frac{1}{2}\right) + \frac{9}{2}. \end{aligned}$$

Since $b, c \in (-2, 1)$, $b + \frac{1}{2}, c + \frac{1}{2} \in \left(-\frac{3}{2}, \frac{3}{2}\right)$. Thus,

$$\left(b + \frac{1}{2}\right)\left(c + \frac{1}{2}\right) \leqslant \left|b + \frac{1}{2}\right| \cdot \left|c + \frac{1}{2}\right| < \frac{9}{4}.$$

So

$$\begin{aligned} f(-2) &= -2\left(b + \frac{1}{2}\right)\left(c + \frac{1}{2}\right) + \frac{9}{2} \\ &> -2 \cdot \frac{9}{4} + \frac{9}{2} = 0, \end{aligned}$$

$$f(1) = bc - b - c + 1 = (1 - b)(1 - c) > 0.$$

Hence, when $x \in (-2, 1)$, $f(x)$ is always greater than 0. Therefore,

$$f(a) > 0,$$

or

$$abc > a + b + c - 2.$$

Eg. 4 Let $x_1, x_2, x_3, y_1, y_2, y_3$ be real numbers such that $x_1^2 + x_2^2 + x_3^2 \leqslant 1$. Prove that

$$(x_1y_1 + x_2y_2 + x_3y_3 - 1)^2 \geqslant (x_1^2 + x_2^2 + x_3^2 - 1)(y_1^2 + y_2^2 + y_3^2 - 1).$$

Proof. When $x_1^2+x_2^2+x_3^2=1$, the original inequality is obviously true.

When $x_1^2+x_2^2+x_3^2<1$, construct a quadratic function:

$$\begin{aligned} f(t) &= (x_1^2+x_2^2+x_3^2-1)t^2-2(x_1y_1+x_2y_2+x_3y_3-1)t \\ &\quad +(y_1^2+y_2^2+y_3^2-1) \\ &= (x_1t-y_1)^2+(x_2t-y_2)^2+(x_3t-y_3)^2-(t-1)^2. \end{aligned}$$

This is a parabola opening downwards. Also,

$$f(1)=(x_1-y_1)^2+(x_2-y_2)^2+(x_3-y_3)^2\geqslant 0.$$

Hence, the parabola must intersect with the x-*axis*. Thus,

$$\begin{aligned} \Delta &= 4(x_1y_1+x_2y_2+x_3y_3-1)^2 \\ &\quad -4(x_1^2+x_2^2+x_3^2-1)(y_1^2+y_2^2+y_3^2-1)\geqslant 0, \end{aligned}$$

Therefore

$$(x_1y_1+x_2y_2+x_3y_3-1)^2\geqslant(x_1^2+x_2^2+x_3^2-1)(y_1^2+y_2^2+y_3^2-1).$$

Note. To prove inequalities like $A\cdot C\geqslant$ (or $\leqslant$)B^2, we first change the inequality into

$$4A\cdot C \text{ (or } \leqslant)(2B)^2,$$

then we can construct a quadratic function $f(x)=Ax^2-(2B)x+C$, and manage to prove that its determinant $\Delta\leqslant 0$ (or $\geqslant 0$).

Eg. 5 Assume $\triangle ABC$ has three sides with length a, b, c that satisfy $a+b+c=1$.

Show that

$$5(a^2+b^2+c^2)+18abc\geqslant\frac{7}{3}.$$

Proof. Since

$$\begin{aligned} a^2+b^2+c^2 &= (a+b+c)^2-2(ab+bc+ca) \\ &= 1-2(ab+bc+ca). \end{aligned}$$

The original inequality is equivalent to

$$\frac{5}{9}(ab+bc+ca)-abc \leqslant \frac{4}{27}. \qquad ①$$

Let $f(x)=(x-a)(x-b)(x-c)=x^3-x^2+(ab+bc+ca)x-abc$, then

$$f\left(\frac{5}{9}\right)=\left(\frac{5}{9}\right)^3-\left(\frac{5}{9}\right)^2+\frac{5}{9}(ab+bc+ca)-abc.$$

Since a, b, c are the three sides of a triangle, we have $a, b, c \in \left(0, \frac{1}{2}\right)$, hence $\frac{5}{9}-a$, $\frac{5}{9}-b$ and $\frac{5}{9}-c$ are all greater than 0. So

$$\begin{aligned} f\left(\frac{5}{9}\right) &= \left(\frac{5}{9}-a\right)\left(\frac{5}{9}-b\right)\left(\frac{5}{9}-c\right) \\ &\leqslant \frac{1}{27}\cdot\left[\left(\frac{5}{9}-a\right)+\left(\frac{5}{9}-b\right)+\left(\frac{5}{9}-c\right)\right]^3 \\ &= \frac{8}{27^2}. \end{aligned}$$

Therefore, $\frac{8}{27^2} \geqslant \left(\frac{5}{9}\right)^3-\left(\frac{5}{9}\right)^2+\frac{5}{9}(ab+bc+ca)-abc.$

Rearrange it and we get ①, so the original inequality is proved.

Eg. 6 Assume the inequality

$$\sqrt{2}(2a+3)\cos\left(\theta-\frac{\pi}{4}\right)+\frac{6}{\sin\theta+\cos\theta}-2\sin 2\theta<3a+6$$

holds for any $\theta \in \left[0, \frac{\pi}{2}\right]$. Find the possible range for a.

Soln. Let $\sin\theta+\cos\theta=x$, then $x\in[1, \sqrt{2}]$, and

$$\sin 2\theta = 2\sin\theta\cos\theta = x^2-1,$$

$$\cos\left(\theta-\frac{\pi}{4}\right)=\frac{\sqrt{2}}{2}\cos\theta+\frac{\sqrt{2}}{2}\sin\theta=\frac{\sqrt{2}}{2}x.$$

Hence the original inequality turns out to be

$$(2a+3)x+\frac{6}{x}-2(x^2-1)<3a+6,$$

or

$$2x^3-(2a+3)x^2+(3a+4)x-6>0,$$
$$(2x-3)(x^2-ax+2)>0.$$

Since $x\in[1,\sqrt{2}]$, $2x-3<0$, thus the inequality

$$x^2-ax+2<0$$

holds for any $x\in[1,\sqrt{2}]$. In other words,

$$a>x+\frac{2}{x},\ x\in[1,\sqrt{2}]$$

is always true.

Denote $f(x)=x+\frac{2}{x}$, $x\in[1,\sqrt{2}]$, then $a>f_{\max}(x)$.

Because $f(x)=x+\frac{2}{x}$ is monotonically decreasing on $[1,\sqrt{2}]$, $f_{\max}(x)=f(1)=3$, therefore $a>3$.

Note. Using the monotonicity of a function as well as finding the extremum of such function can help us solving the inequality.

Eg. 7 Prove that for any positive real numbers a, b, c, we have

$$1<\frac{a}{\sqrt{a^2+b^2}}+\frac{b}{\sqrt{b^2+c^2}}+\frac{c}{\sqrt{c^2+a^2}}\leqslant\frac{3\sqrt{2}}{2}.$$

Proof. Let $x=\frac{b^2}{a^2}$, $y=\frac{c^2}{b^2}$, $z=\frac{a^2}{c^2}$.

Then $x, y, z\in\mathbf{R}^+$ and $xyz=1$. We only need to show that

$$1<\frac{1}{\sqrt{1+x}}+\frac{1}{\sqrt{1+y}}+\frac{1}{\sqrt{1+z}}\leqslant\frac{3\sqrt{2}}{2}.$$

WLOG, assume that $x\leqslant y\leqslant z$. Denote $A=xy$, then $z=\frac{1}{A}$ and $A\leqslant 1$. Hence,

$$\frac{1}{\sqrt{1+x}}+\frac{1}{\sqrt{1+y}}+\frac{1}{\sqrt{1+z}}>\frac{1}{\sqrt{1+x}}+\frac{1}{\sqrt{1+\frac{1}{x}}}$$
$$=\frac{1+\sqrt{x}}{\sqrt{1+x}}>1.$$

Now let $u = \dfrac{1}{\sqrt{1 + A + x + \dfrac{A}{x}}}$, then $u \in \left(0, \dfrac{1}{1+\sqrt{A}}\right]$ and $u = \dfrac{1}{1+\sqrt{A}}$ if and only if $x = \sqrt{A}$. Thus,

$$\begin{aligned}
&\left(\frac{1}{\sqrt{1+x}} + \frac{1}{\sqrt{1+y}}\right)^2 \\
&= \left(\frac{1}{\sqrt{1+x}} + \frac{1}{\sqrt{1+\dfrac{A}{x}}}\right)^2 \\
&= \frac{1}{1+x} + \frac{1}{1+\dfrac{A}{x}} + \frac{2}{\sqrt{1+A+x+\dfrac{A}{x}}} \\
&= \frac{2 + x + \dfrac{A}{x}}{1 + A + x + \dfrac{A}{x}} + \frac{2}{\sqrt{1+A+x+\dfrac{A}{x}}} \\
&= 1 + (1-A)u^2 + 2u.
\end{aligned}$$

Construct the function $f(u) = (1-A)u^2 + 2u + 1$, then $f(u)$ is increasing for $u \in \left(0, \dfrac{1}{1+\sqrt{A}}\right]$. So

$$\frac{1}{\sqrt{1+x}} + \frac{1}{\sqrt{1+y}} \leqslant \sqrt{f\left(\frac{1}{1+\sqrt{A}}\right)} = \frac{2}{\sqrt{1+\sqrt{A}}}.$$

Let $\sqrt{A} = v$, then

$$\begin{aligned}
&\frac{1}{\sqrt{1+x}} + \frac{1}{\sqrt{1+y}} + \frac{1}{\sqrt{1+z}} \\
&\leqslant \frac{2}{\sqrt{1+\sqrt{A}}} + \frac{1}{\sqrt{1+\dfrac{1}{A}}} \\
&= \frac{2}{\sqrt{1+v}} + \frac{\sqrt{2}v}{\sqrt{2(1+v^2)}} \\
&\leqslant \frac{2}{\sqrt{1+v}} + \frac{\sqrt{2}v}{1+v}
\end{aligned}$$

$$= \frac{2}{\sqrt{1+v}} + \sqrt{2} - \frac{\sqrt{2}}{1+v}$$

$$= -\sqrt{2}\left(\frac{1}{\sqrt{1+v}} - \frac{\sqrt{2}}{2}\right)^2 + \frac{3\sqrt{2}}{2}$$

$$\leqslant \frac{3\sqrt{2}}{2}.$$

5.3 Construct Graph

If the quantities in the problem has geometric interpretations, we can then plot a related graph and visualize these quantities. Eventually, we can verify the conclusion from the graph we constructed.

Eg. 8 Prove that for any real number x, we have

$$\left|\sqrt{x^2+x+1} - \sqrt{x^2-x+1}\right| < 1.$$

Proof. Since

$$\left|\sqrt{x^2+x+1} - \sqrt{x^2-x+1}\right|$$
$$= \left|\sqrt{\left(x+\frac{1}{2}\right)^2 + \left(\frac{\sqrt{3}}{2}\right)^2} - \sqrt{\left(x-\frac{1}{2}\right)^2 + \left(\frac{\sqrt{3}}{2}\right)^2}\right|.$$

We see the above as the difference of two distances in the rectangular coordinates system: the distance from $P\left(x, \frac{\sqrt{3}}{2}\right)$ to $A\left(-\frac{1}{2}, 0\right)$ and the distance from P to $B\left(\frac{1}{2}, 0\right)$.

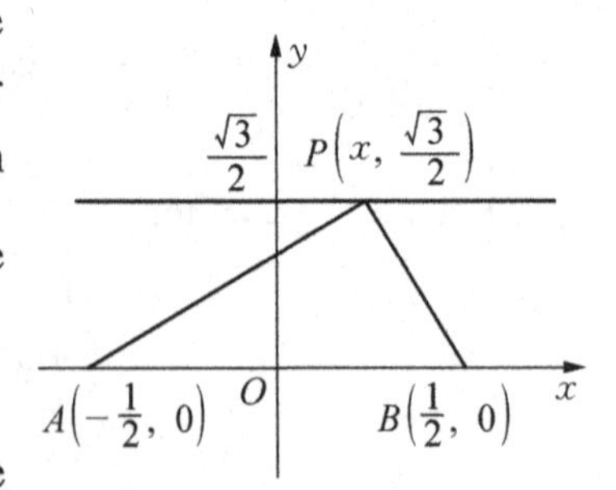

Figure 5.1

Since for a triangle, the difference of two sides is always smaller than the third side, thus

$$\left|\sqrt{x^2+x+1} - \sqrt{x^2-x+1}\right| < AB = 1.$$

Eg. 9 Assume x, y, z are real numbers, $0 < x < y < z < \frac{\pi}{2}$. Prove that

$$\frac{\pi}{2} + 2\sin x\cos y + 2\sin y\cos z > \sin 2x + \sin 2y + \sin 2z.$$

Proof. The original inequality is equivalent to

$$\frac{\pi}{4} > \sin x(\cos x - \cos y) + \sin y(\cos y - \cos z) + \sin z\cos z.$$

Construct a graph, as shown in Figure 5.2:

Circle O is a unit circle, S_1, S_2, S_3 represent the area of the three rectangular respectively. Thus,

$$S_1 = \sin x(\cos x - \cos y),$$
$$S_2 = \sin y(\cos y - \cos z),$$
$$S_3 = \sin z\cos z.$$

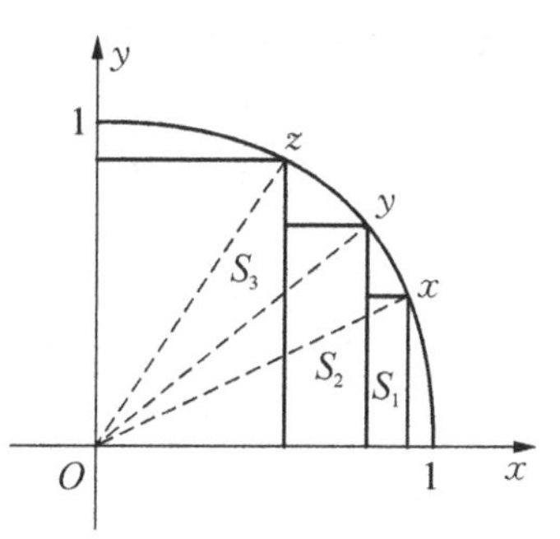

Figure 5.2

Since $S_1 + S_2 + S_3 < \frac{1}{4} \cdot \pi \cdot 1^2 = \frac{1}{4}\pi$, we have

$$\frac{\pi}{4} > \sin x(\cos x - \cos y) + \sin y(\cos y - \cos z) + \sin z\cos z.$$

Therefore, the original inequality holds.

Eg. 10 Let x, y, z, α, β, γ be positive real numbers. α, β, $\gamma \in [0, \pi)$ and among them, the sum of any two is greater than the third. Show that

$$\sqrt{x^2 + y^2 - 2xy\cos\alpha} + \sqrt{y^2 + z^2 - 2yz\cos\beta} \geqslant \sqrt{z^2 + x^2 - 2zx\cos\gamma}.$$

Hint. The terms in the inequality reminds us of the law of cosines, hence we should construct some triangles.

Proof. Since $\alpha < \beta + \gamma < \pi$, $\beta < \gamma + \alpha < \pi$, $\gamma < \alpha + \beta < \pi$, from a point P in space, we can construct a trihedral angle P-ABC, such that:

$$\angle APB = \alpha,$$
$$\angle BPC = \beta,$$
$$\angle CPA = \gamma,$$

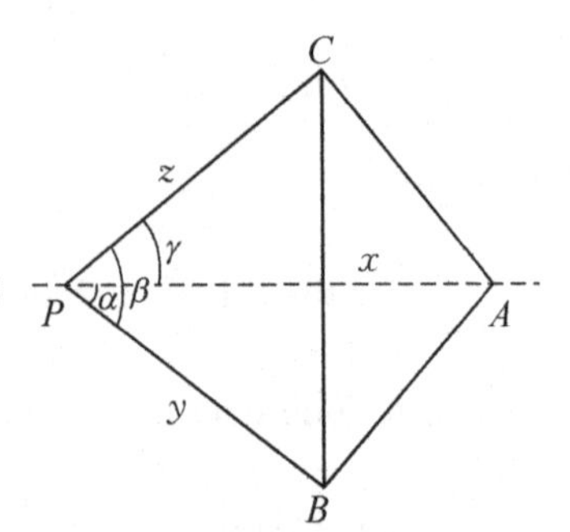

Figure 5.3

and that $PA = x$, $PB = y$, $PC = z$, as shown in Figure 5.3.

As a result, since $AB + BC \geqslant AC$, the original inequality holds true.

Eg. 11 Given that $|u| \leqslant \sqrt{2}$, v is a positive real number. Prove that

$$S = (u-v)^2 + \left(\sqrt{2-u^2} - \frac{9}{v}\right)^2 \geqslant 8.$$

Proof. The key of this problem is to view S as the squared distance between point $A(u, \sqrt{2-u^2})$ and $B\left(u, \frac{9}{v}\right)$ in the rectangular coordinates system.

Obviously, point A is on the circle $x^2 + y^2 = 2$, point B is on the hyperbola $xy = 9$.

As a result, we need to find the minimum distance between the circle $x^2 + y^2 = 2$ and the hyperbola $xy = 9$.

On the graph it is easy to see that the minimum distance is from A $(1, 1)$ to B $(3, 3)$ which is $2\sqrt{2}$. Therefore the minimum for S is $(2\sqrt{2})^2 = 8$. So $S \geqslant 8$.

5.4 Construct Antithetic Formula

When dealing with an inequality with rotating symmetry, constructing an antithetic formula and combining it with the original inequality may be quite effective.

Eg. 12 Let $a_1, a_2, \cdots, a_n$ be real numbers with sum 1. Prove that

$$\frac{a_1^4}{a_1^3 + a_1^2 a_2 + a_1 a_2^2 + a_2^3} + \frac{a_2^4}{a_2^3 + a_2^2 a_3 + a_2 a_3^2 + a_3^3} + \cdots$$

$$+\frac{a_n^4}{a_n^3+a_n^2a_1+a_na_1^2+a_1^3}\geqslant\frac{1}{4}.$$

Proof. Denote A the left hand side of the original inequality. Construct an antithetic formula

$$B=\frac{a_2^4}{a_1^3+a_1^2a_2+a_1a_2^2+a_2^3}+\cdots+\frac{a_1^4}{a_n^3+a_n^2a_1+a_na_1^2+a_1^3}.$$

Hence,

$$\begin{aligned}A-B&=\frac{(a_1^2+a_2^2)(a_1+a_2)(a_1-a_2)}{(a_1^2+a_2^2)(a_1+a_2)}+\cdots\\&\quad+\frac{(a_n^2+a_1^2)(a_n+a_1)(a_n-a_1)}{(a_n^2+a_1^2)(a_n+a_1)}\\&=(a_1-a_2)+(a_2-a_3)+\cdots+(a_n-a_1)\\&=0.\end{aligned}$$

Thus $A=B$.

On the other hand, since

$$\begin{aligned}\frac{a_1^4+a_2^4}{(a_1^2+a_2^2)(a_1+a_2)}&\geqslant\frac{(a_1^2+a_2^2)^2}{2(a_1^2+a_2^2)(a_1+a_2)}\\&=\frac{a_1^2+a_2^2}{2(a_1+a_2)}\geqslant\frac{(a_1+a_2)^2}{4(a_1+a_2)}\\&=\frac{a_1+a_2}{4},\end{aligned}$$

so

$$\begin{aligned}A&=\frac{1}{2}(A+B)\\&\geqslant\frac{1}{2}\left[\frac{1}{4}(a_1+a_2)+\frac{1}{4}(a_2+a_3)+\cdots+\frac{1}{4}(a_n+a_1)\right]\\&=\frac{1}{4}.\end{aligned}$$

5.5 Construct Sequence

When we handle an inequality related with n, we can consider constructing a sequence and using the sequence properties (such as monotonicity) to solve the problem.

Eg. 13 Let $x_n = \sqrt{2+\sqrt[3]{3+\cdots+\sqrt[n]{n}}}$, $n \in \mathbf{N}_+$. Prove that

$$x_{n+1} - x_n < \frac{1}{n!}, \ n = 2, 3, \cdots.$$

Proof. When $n = 2$, $x_3 - x_2 = \sqrt{2+\sqrt[3]{3}} - \sqrt{2} < \frac{1}{2!}$.

When $n \geqslant 3$, construct sequences $\{a_i\}$, $\{b_i\}$, $\{c_i\}$ as follows:

$$a_i = \sqrt[i]{i+\sqrt[i+1]{(i+1)+\cdots+\sqrt[n]{n+\sqrt[n+1]{n+1}}}}, \ i = 2, \cdots, n+1.$$

$$b_i = \sqrt[i]{i+\sqrt[i+1]{(i+1)+\cdots+\sqrt[n]{n}}}, \ i = 2, 3, \cdots, n, \ b_{n+1} = 0.$$

$$c_i = a_i^{i-1} + a_i^{i-2}b_i + \cdots + a_i b_i^{i-2} + b_i^{i-1}, \ i = 2, 3, \cdots.$$

Clearly, $x_{n+1} = a_2$, $x_n = b_2$, and

$$(a_i - b_i)c_i = a_i^i - b_i^i = a_{i+1} - b_{i+1}.$$

Hence,

$$a_i - b_i = \frac{a_{i+1} - b_{i+1}}{c_i}, \ i = 2, 3, \cdots, n.$$

Multiply the above $n-1$ equations, and notice that

$$a_{n+1} - b_{n+1} = (n+1)^{\frac{1}{n+1}},$$

we obtain

$$a_2 - b_2 = \frac{a_{n+1} - b_{n+1}}{c_2 c_3 \cdots c_n} = \frac{(n+1)^{\frac{1}{n+1}}}{c_2 c_3 \cdots c_n}.$$

Moreover, since $a_k > b_k \geqslant \sqrt[k]{k}$, we have $c_k \geqslant k \cdot k^{\frac{k-1}{k}} > k \cdot k^{\frac{k-1}{k+1}}$. As a result,

$$x_{n+1} - x_n = a_2 - b_2 < \frac{1}{n!} \cdot \frac{(n+1)^{\frac{1}{n+1}}}{n^{\frac{n-1}{n+1}}} < \frac{1}{n!}.$$

The final step is true because when $n > 2$, $\frac{n+1}{n^{n-1}} < \frac{2n}{n^2} < 1$.

5.6 Construct Auxiliary Statement

Sometimes if it is hard to prove a statement directly, we can try to form an auxiliary statement to assist proving the original statement.

Eg. 14 Given two sequences $x_1, x_2, \cdots, x_n$ and $y_1, y_2, \cdots, y_n$. We know that

$$x_1 > x_2 > \cdots > x_n > 0, \ y_1 > y_2 > \cdots > y_n > 0,$$

$$x_1 > y_1, \ x_1 + x_2 > y_1 + y_2, \ \cdots,$$

$$x_1 + x_2 + \cdots + x_n > y_1 + y_2 + \cdots + y_n.$$

Prove that for any positive integer k, we have

$$x_1^k + x_2^k + \cdots + x_n^k > y_1^k + y_2^k + \cdots + y_n^k.$$

Hint. We guess whether the following could be true:

$$x_1^k + x_2^k + \cdots + x_n^k > x_1^{k-1}y_1 + x_2^{k-1}y_2 + \cdots + x_n^{k-1}y_n,$$

$$x_1^{k-1}y_1 + x_2^{k-1}y_2 + \cdots + x_n^{k-1}y_n > x_1^{k-2}y_1^2 + x_2^{k-2}y_2^2 + \cdots + x_n^{k-2}y_n^2,$$

$$\cdots\cdots$$

$$x_1y_1^{k-1} + x_2y_2^{k-1} + \cdots + x_ny_n^{k-1} > y_1^k + y_2^k + \cdots + y_n^k.$$

Hence we come up with the following auxiliary statement: If $a_1 > a_2 > \cdots > a_n > 0$ and satisfy the conditions, then

$$a_1x_1 + a_2x_2 + \cdots + a_nx_n > a_1y_1 + a_2y_2 + \cdots + a_ny_n. \quad ①$$

Proof. Since $a_1 > a_2 > \cdots > a_n > 0$, there exist positive numbers $b_1, b_2, \cdots, b_n$, such that

$$\begin{aligned} a_n &= b_1, \\ a_{n-1} &= b_1 + b_2, \\ &\cdots\cdots \\ a_2 &= b_1 + b_2 + \cdots + b_{n-1}, \\ a_1 &= b_1 + b_2 + \cdots + b_n. \end{aligned}$$

Thus,

$$\begin{aligned}&a_1x_1+a_2x_2+\cdots+a_nx_n\\&=(b_1+b_2+\cdots+b_n)x_1+(b_1+b_2+\cdots+b_{n-1})x_2+\cdots+b_1x_n\\&=b_1(x_1+x_2+\cdots+x_n)+b_2(x_1+x_2+\cdots+x_{n-1})+\cdots+b_nx_1\\&>b_1(y_1+y_2+\cdots+y_n)+b_2(y_1+y_2+\cdots+y_{n-1})+\cdots+b_ny_1\\&=(b_1+b_2+\cdots+b_n)y_1+(b_1+b_2+\cdots+b_{n-1})y_2+\cdots+b_1y_n\\&=a_1y_1+a_2y_2+\cdots+a_ny_n.\end{aligned}$$

Hence ① holds.

Now, let $a_i=x_i^{k-1}$, $x_i^{k-2}y_i$, $\cdots$, $y_i^{k-1}(i=1,2,\cdots,n)$ in order, it is then easy to see that the original inequality holds.

5.7 Construct Example or Counter Example

Eg. 15 For a given integer $n>1$, do there exist $2n$ mutually different positive integers $a_1, a_2, \cdots, a_n$ and $b_1, b_2, \cdots, b_n$ such that the following conditions hold:

(1) $a_1+a_2+\cdots+a_n=b_1+b_2+\cdots+b_n$.

(2) $n-1>\sum\limits_{i=1}^{n}\frac{a_i-b_i}{a_i+b_i}>n-1-\frac{1}{2002}$.

Soln. The answer is affirmative.

Pick

$$a_1=N+1,\ a_2=N+2,\ \cdots,\ a_{n-1}=N+(n-1),$$
$$b_1=1,\ b_2=2,\ \cdots,\ b_{n-1}=n-1.$$

Then, from (1) we have $b_n-a_n=N(n-1)$. Set $a_n=N^2$, $b_n=N^2+N(n-1)$. Thus,

$$\begin{aligned}\sum_{i=1}^{n}\frac{a_i-b_i}{a_i+b_i}&=n-1-\left(\frac{2}{N+2}+\cdots+\frac{2(n-1)}{2(n-1)+N}+\frac{n-1}{2N+(n-1)}\right)\\&<n-1.\end{aligned}$$

(We have used $\frac{a_i-b_i}{a_i+b_i}=1-\frac{2i}{2i+N}$, $1\leqslant i\leqslant n-1$ in the above formula.)

Now pick $N>2002n(n-1)$, then

$$\sum_{i=1}^{n} \frac{a_i - b_i}{a_i + b_i} > n - 1 - \frac{2n(n-1)}{N} > n - 1 - \frac{1}{2002}.$$

Eg. 16 Given positive integer $n \geqslant 2$, real numbers $a_1 \geqslant a_2 \geqslant \cdots \geqslant a_n > 0$, $b_1 \geqslant b_2 \geqslant \cdots \geqslant b_n > 0$ satisfy $a_1 a_2 \cdots a_n = b_1 b_2 \cdots b_n$ and $\sum_{1 \leqslant i < j \leqslant n} (a_i - a_j) \leqslant \sum_{1 \leqslant i < j \leqslant n} (b_i - b_j)$. Given these conditions, does $\sum_{i=1}^{n} a_i \leqslant (n-2) \sum_{i=1}^{n} b_i$ always hold?

Soln. Not really.

Set

$$a_1 = a_2 = \cdots = a_{n-1} = h, \ a_n = \frac{1}{h^{n-1}},$$

$$b_1 = k, \ b_2 = b_3 = \cdots = b_{n-1} = 1, \ b_n = \frac{1}{k} \ (k \geqslant 1).$$

(Our idea is to pick h large enough so that $\sum_{i=1}^{n} a_i$ is large enough. In order not to make $\sum_{1 \leqslant i < j \leqslant n} (a_i - a_j)$ too large, we can pick $a_1, a_2, \cdots, a_{n-1}$ to be the same and we consider choosing of b_i's in the reverse direction.)

Now, we see that $\sum_{1 \leqslant i < j \leqslant n} (a_i - b_j) \leqslant \sum_{1 \leqslant i < j \leqslant n} (b_i - b_j)$ is equivalent to

$$(n-1)h - \frac{n-1}{h^{n-1}} \leqslant (n-1)k - \frac{n-1}{k},$$

or

$$k - h \geqslant \frac{1}{k} - \frac{1}{h^{n-1}}.$$

Note that $k \leqslant 1$, $h > 0$, so we only require $k - h \geqslant 1$, or $k \geqslant h + 1$. On the other hand,

$$\sum_{i=1}^{n} a_i > (n-2) \sum_{i=1}^{n} b_i$$

is equivalent to

$$(n-1)h+\frac{1}{h^{n-1}}>(n-2)\left[k+(n-2)+\frac{1}{k}\right].$$

For simplicity, we pick $k=h+1$, so we need

$$(n-1)h>(n-2)[h+1+(n-2)+1],$$

or $h>n^2-2n$. We can thus pick $h=n^2-2n+1=(n-1)^2$, then we will have $\sum_{i=1}^{n}a_i>(n-2)\sum_{i=1}^{n}b_i$.

Note. In fact, we can prove that $\sum_{i=1}^{n}a_i\leqslant(n-1)\sum_{i=1}^{n}b_i$, please refer to Exercise 1 Problem 15.

Exercise 5

1. Assume that $k>a>b>c>0$, show that

$$k^2-(a+b+c)k+(ab+bc+ca)>0.$$

2. Assume that the three roots α, β, γ of the equation $x^3+ax^2+bx+c=0$ are all real numbers, and $a^2=2b+2$. Prove that $|a-c|\leqslant 2$.

3. Let $F(x)=|f(x)\cdot g(x)|$, where $f(x)=ax^2+bx+c$, $x\in[-1,1]$ and $g(x)=cx^2+bx+a$, $x\in[-1,1]$. Also, for any parameters a, b, c, $|f(x)|\leqslant 1$ always holds. Find the maximum for $F(x)$.

4. Let a, b, c be real numbers, prove that

$$\frac{|a+b+c|}{1+|a+b+c|}\leqslant\frac{|a|}{1+|a|}+\frac{|b|}{1+|b|}+\frac{|c|}{1+|c|}.$$

5. Find all real numbers a, such that any positive integer solutions of the inequality $x^2+y^2+z^2\leqslant a(xy+yz+zx)$ are three sides of some triangle.

6. Let a_1, a_2, a_3, b_1, b_2, b_3 be positive real numbers. Show that

$$(a_1b_2 + a_2b_1 + a_2b_3 + a_3b_2 + a_3b_1 + a_1b_3)^2$$
$$\geqslant 4(a_1a_2 + a_2a_3 + a_3a_1)(b_1b_2 + b_2b_3 + b_3b_1).$$

The equal sign holds if and only if $\frac{a_1}{b_1} = \frac{a_2}{b_2} = \frac{a_3}{b_3}$.

7. Assume $\{a_n\} \in \mathbf{R}$ satisfy $a_{n+1} \geqslant a_n^2 + \frac{1}{5}$, $n \geqslant 0$. Prove that

$$\sqrt{a_{n+5}} \geqslant a_{n-5}^2, \; n \geqslant 5.$$

8. Let x, y, z be real numbers. Prove that

$$\sqrt{x^2 + xy + y^2} + \sqrt{x^2 + xz + z^2} \geqslant \sqrt{y^2 + yz + z^2}.$$

9. Let x, y, z be positive real numbers, prove that

$$3\sqrt{xy + yz + zx}$$
$$\leqslant \sqrt{x^2 + xy + y^2} + \sqrt{y^2 + yz + z^2} + \sqrt{z^2 + zx + x^2}$$
$$\leqslant 2(x + y + z).$$

10. Assume α, β, γ are all acute angles, and $\cos^2\alpha + \cos^2\beta + \cos^2\gamma = 1$. Prove that

$$\frac{3\pi}{4} < \alpha + \beta + \gamma < \pi.$$

11. If p, q are real numbers, and for $0 \leqslant x \leqslant 1$, we have the following inequality:

$$\left| \sqrt{1 - x^2} - px - q \right| \leqslant \frac{\sqrt{2} - 1}{2}.$$

Prove that $p = -1$, $q = \frac{1 + \sqrt{2}}{2}$.

12. (Minkovski Inequality) Prove that, for any $2n$ positive numbers $a_1, a_2, \cdots, a_n$ and $b_1, b_2, \cdots, b_n$, we have

$$\sqrt{a_1^2 + b_1^2} + \sqrt{a_2^2 + b_2^2} + \cdots + \sqrt{a_n^2 + b_n^2}$$
$$\geqslant \sqrt{(a_1 + a_2 + \cdots + a_n)^2 + (b_1 + b_2 + \cdots + b_n)^2},$$

and the equality holds if and only if $\frac{b_1}{a_1} = \frac{b_2}{a_2} = \cdots = \frac{b_n}{a_n}$.

13. Assume $0 < a_i \leqslant a (i = 1, 2, \cdots, 6)$. Prove that

(1) $$\frac{\sum_{i=1}^{4} a_i}{a} - \frac{a_1a_2 + a_2a_3 + a_3a_4 + a_4a_1}{a^2} \leqslant 2.$$

(2) $$\frac{\sum_{i=1}^{6} a_i}{a} - \frac{a_1a_2 + a_2a_3 + \cdots + a_6a_1}{a^2} \leqslant 3.$$

14. Assume that 100 positive numbers $x_1, x_2, \cdots, x_{100}$ satisfy

(1) $x_1^2 + x_2^2 + \cdots + x_{100}^2 > 10000$.

(2) $x_1 + x_2 + \cdots + x_{100} \leqslant 300$.

Prove that we can find 3 numbers among them, such that their sum is greater than 100.

15. Assume n is an odd number greater than 2. Prove that only when $n = 3$ or 5, for any $a_1, a_2, \cdots, a_n \in \mathbf{R}$ the following inequality is true:

$$(a_1 - a_2)(a_1 - a_3)\cdots(a_1 - a_n) + (a_2 - a_1)(a_2 - a_3)\cdots(a_2 - a_n) + \cdots + (a_n - a_1)(a_n - a_2)\cdots(a_n - a_{n-1}) \geqslant 0.$$

16. Assume $a_1, a_2, \cdots, a_n (n \geqslant 2)$ are all greater than -1 and are of the same sign. Prove that

$$(1 + a_1)(1 + a_2)\cdots(1 + a_n) > 1 + a_1 + a_2 + \cdots + a_n.$$

17. Assume a_i are positive real numbers $(i = 1, 2, \cdots, n)$. Denote

$$kb_k = a_1 + a_2 + \cdots + a_k (k = 1, 2, \cdots, n),$$

$$C_n = (a_1 - b_1)^2 + (a_2 - b_2)^2 + \cdots + (a_n - b_n)^2,$$

$$D_n = (a_1 - b_n)^2 + (a_2 - b_n)^2 + \cdots + (a_n - b_n)^2.$$

Prove that $C_n \leqslant D_n \leqslant 2C_n$.

Chapter 6 Local Inequality

For an inequality that appears in sum form, if we find it hard to handle the inequality directly, we can estimate some local properties first and obtain some local inequalities. Combining these local inequalities, we might be able to deduct the original inequality. Note that by local, we can mean a single term, or even a combination of several terms.

Eg. 1 Let a, b, c, d be positive real numbers, prove that

$$\frac{a^3+b^3+c^3}{a+b+c}+\frac{b^3+c^3+d^3}{b+c+d}+\frac{c^3+d^3+a^3}{c+d+a}+\frac{d^3+a^3+b^3}{d+a+b}$$
$$\geqslant a^2+b^2+c^2+d^2.$$

Hint. It might be tough to prove the inequality directly. However, if we can prove a local inequality

$$\frac{a^3+b^3+c^3}{a+b+c}\geqslant\frac{a^2+b^2+c^2}{3}. \quad ①$$

Then similarly, we can show $\frac{b^3+c^3+d^3}{b+c+d}\geqslant\frac{b^2+c^2+d^2}{3}$, $\frac{c^3+d^3+a^3}{c+d+a}\geqslant\frac{c^2+d^2+a^2}{3}$, $\frac{d^3+a^3+b^3}{d+a+b}\geqslant\frac{d^2+a^2+b^2}{3}$. Adding them together, we get the desired inequality. In conclusion, to prove the original inequality, we only need to prove ①.

Proof. Let us prove that, for any x, y, $z\in\mathbf{R}^+$,

$$\frac{x^3+y^3+z^3}{x+y+z}\geqslant\frac{x^2+y^2+z^2}{3}.$$

Actually, according to the Cauchy-Schwarz Inequality,

$$(x+y+z)(x^3+y^3+z^3) \geqslant (x^2+y^2+z^2)^2$$
$$\geqslant (x^2+y^2+z^2) \cdot \frac{(x+y+z)^2}{3}.$$

Hence,

$$\frac{x^3+y^3+z^3}{x+y+z} \geqslant \frac{x^2+y^2+z^2}{3}.$$

From what we have discussed before, the original inequality holds.

Eg. 2 Let x, y, z be positive real numbers, prove that

$$\sqrt{\frac{x}{y+z}}+\sqrt{\frac{y}{z+x}}+\sqrt{\frac{z}{x+y}} \geqslant 2.$$

Hint. It seems quite hard to directly handle the left hand side of the inequality.

So we start with a local inequality and show that

$$\sqrt{\frac{x}{y+z}} \geqslant \frac{2x}{x+y+z}.$$

Proof. First we prove that

$$\sqrt{\frac{x}{y+z}} \geqslant \frac{2x}{x+y+z}.$$

In fact, using the AM-GM Inequality, we have

$$\sqrt{\frac{x}{y+z}} = \frac{x}{\sqrt{x}\sqrt{y+z}} \geqslant \frac{x}{\frac{x+y+z}{2}} = \frac{2x}{x+y+z}.$$

Similarly,

$$\sqrt{\frac{y}{z+x}} \geqslant \frac{2y}{x+y+z},$$
$$\sqrt{\frac{z}{x+y}} \geqslant \frac{2z}{x+y+z}.$$

Adding the above three inequalities, we obtain

$$\sqrt{\frac{x}{y+z}}+\sqrt{\frac{y}{z+x}}+\sqrt{\frac{z}{x+y}}\geqslant 2.$$

Note. The equal sign in this inequality can never be reached. Why?

Eg. 3 Assume that $0\leqslant a, b, c\leqslant 1$, prove that

$$\frac{a}{bc+1}+\frac{b}{ca+1}+\frac{c}{ab+1}\leqslant 2.$$

Hint. We find that the equal sign is not reached when $a=b=c$; instead, it is reached when there are one 0 and two 1s in a, b, c. Knowing this, we shouldn't consider the inequality as a whole and we should start with investigating single terms.

Proof. First, we prove

$$\frac{a}{bc+1}\leqslant\frac{2a}{a+b+c}. \quad ①$$

Note that ① is equivalent to

$$a+b+c\leqslant 2bc+2,$$

or

$$(b-1)(c-1)+bc+1\geqslant a.$$

Since $a, b, c\in[0, 1]$, the above inequality is obviously true, so ① holds.

Similarly, we have

$$\frac{b}{ca+1}\leqslant\frac{2b}{a+b+c}, \quad ②$$

$$\frac{c}{ab+1}\leqslant\frac{2c}{a+b+c}. \quad ③$$

Adding ①, ② and ③ together, the original inequality is proved.

Eg. 4 Assume that $x_i\geqslant 1$, $i=1, 2, \cdots, n$ and $x_1x_2\cdots x_n=a^n$. Denote $x_{n+1}=x_1$.

Prove that when $n>2$, we have

$$\sum_{i=1}^{n} x_i x_{i+1} - \sum_{i=1}^{n} x_i \geqslant \frac{n}{2}(a^2 - 1).$$

Hint. To relate $x_i x_{i+1}$ with x_i, naturally we think of $(x_i - 1)(x_{i+1} - 1)$. Based on the given conditions, $(x_i - 1)(x_{i+1} - 1) \geqslant 0$.

Proof. From the hint, we have

$$(x_i - 1)(x_{i+1} - 1) \geqslant 0,$$

thus

$$2x_i x_{i+1} - x_i - x_{i+1} \geqslant x_i x_{i+1} - 1.$$

Sum the above inequality for $i = 1, 2, \cdots, n$, we obtain

$$2\left(\sum_{i=1}^{n} x_i x_{i+1} - \sum_{i=1}^{n} x_i\right) \geqslant \sum_{i=1}^{n} x_i x_{i+1} - n.$$

Using the AM-GM inequality, we have

$$\sum_{i=1}^{n} x_i x_{i+1} \geqslant n \cdot \sqrt[n]{(a^n)^2} = na^2.$$

Therefore,

$$2\left(\sum_{i=1}^{n} x_i x_{i+1} - \sum_{i=1}^{n} x_i\right) \geqslant n(a^2 - 1),$$

the original inequality is true.

Eg. 5 Assume real numbers $a_1, a_2, \cdots, a_n \in (-1, 1]$, prove that

$$\sum_{i=1}^{n} \frac{1}{1 + a_i a_{i+1}} \geqslant \sum_{i=1}^{n} \frac{1}{1 + a_i^2} \text{ (Denote } a_{n+1} = a_1\text{)}.$$

Proof. First let us prove that if $x, y \in (-1, 1]$, then

$$\frac{2}{1 + xy} \geqslant \frac{1}{1 + x^2} + \frac{1}{1 + y^2}, \qquad ①$$

① is equivalent to

$$2(1 + x^2)(1 + y^2) - (1 + xy)(2 + x^2 + y^2) \geqslant 0,$$

or

$$(x-y)^2 - xy(x-y)^2 \geqslant 0.$$

Hence ① holds.

As a result,

$$\frac{2}{1+a_1a_2} \geqslant \frac{1}{1+a_1^2} + \frac{1}{1+a_2^2},$$

$$\frac{2}{1+a_2a_3} \geqslant \frac{1}{1+a_2^2} + \frac{1}{1+a_3^2},$$

$$\cdots\cdots$$

$$\frac{2}{1+a_na_1} \geqslant \frac{1}{1+a_n^2} + \frac{1}{1+a_1^2}.$$

Summing the above n inequalities and we get the original inequality.

Eg. 6 Let x, y, z be non-negative real numbers satisfying $x^2 + y^2 + z^2 = 1$. Prove that

$$\frac{x}{1+yz} + \frac{y}{1+zx} + \frac{z}{1+xy} \geqslant 1.$$

Proof. We only need to prove the local inequality:

$$\frac{x}{1+yz} \geqslant x^2, \qquad ①$$

① is equivalent to

$$x + xyz \leqslant 1.$$

However,

$$\begin{aligned} x + xyz &\leqslant x + \frac{1}{2}x(y^2+z^2) \\ &= \frac{1}{2}(3x - x^3) \\ &= \frac{1}{2}[2-(x-1)^2(x+2)] \\ &\leqslant 1. \end{aligned}$$

Hence ① holds, therefore the original inequality is true.

Note. We can prove the following chain of inequalities:

$$1 \leqslant \frac{x}{1+yz} + \frac{y}{1+zx} + \frac{z}{1+xy}$$

$$\leqslant \frac{x}{1-yz} + \frac{y}{1-zx} + \frac{z}{1-xy}$$

$$\leqslant \frac{3\sqrt{3}}{2}.$$

Eg. 7 Let x, y, z be positive real numbers. Prove that

$$\frac{xz}{x^2+xz+yz} + \frac{xy}{y^2+xy+xz} + \frac{yz}{z^2+yz+xy} \leqslant 1.$$

Proof. WLOG, assume that $xyz = 1$.

First, we prove

$$\frac{xz}{x^2+xz+yz} \leqslant \frac{1}{1+y+\frac{1}{z}}, \qquad ①$$

① is equivalent to

$$x^2+xz+yz \geqslant xz+1+x,$$

or

$$x^2+\frac{1}{x} \geqslant 1+x,$$

or

$$(x-1)^2(x+1) \geqslant 0.$$

Hence ① holds.

Also,

$$\frac{1}{1+y+\frac{1}{z}} = \frac{z}{yz+z+1},$$

and similarly,

$$\frac{xy}{y^2+xy+xz} \leqslant \frac{1}{1+z+\frac{1}{x}} = \frac{1}{1+z+yz},$$

$$\frac{yz}{z^2+yz+xy} \leqslant \frac{1}{1+x+\frac{1}{y}} = \frac{xyz}{xyz+x+xz} = \frac{yz}{1+z+yz}.$$

Therefore,

$$\sum \frac{xz}{x^2+xz+yz} \leqslant \frac{1+z+yz}{1+z+yz} = 1,$$

the original inequality is proved.

Eg. 8 Assume $n(\geqslant 3)$ is an integer, prove that for real numbers $x_1 \leqslant x_2 \leqslant \cdots \leqslant x_n$, we have

$$\frac{x_n x_1}{x_2} + \frac{x_1 x_2}{x_3} + \cdots + \frac{x_{n-1} x_n}{x_1} \geqslant x_1 + x_2 + \cdots + x_n.$$

Proof. First let us prove a lemma: If $0 < x \leqslant y$, $0 < a \leqslant 1$, then

$$x + y \leqslant ax + \frac{y}{a}. \qquad ①$$

In fact, since $ax \leqslant x \leqslant y$, we have $(1-a)(y-ax) \geqslant 0$, or

$$a^2 x + y \geqslant ax + ay.$$

Thus,

$$x + y \leqslant ax + \frac{y}{a}.$$

Now, let $(x, y, a) = \left(x_i, x_{n-1} \cdot \frac{x_{i+1}}{x_2}, \frac{x_{i+1}}{x_{i+2}}\right)$, $i = 1, 2, \cdots, n-2$.
Plugging into ①, we have

$$x_i + \frac{x_{n-1} x_{i+1}}{x_2} \leqslant \frac{x_i x_{i+1}}{x_{i+2}} + x_{n-1} \cdot \frac{x_{i+2}}{x_2}. \qquad ②$$

Sum the inequality ② for $i = 1, 2, \cdots, n-2$, we obtain

$$\begin{aligned} & x_1 + x_2 + \cdots + x_{n-2} + \frac{x_{n-1}}{x_2}(x_2 + x_3 + \cdots + x_{n-1}) \\ \leqslant & \frac{x_1 x_2}{x_3} + \frac{x_2 x_3}{x_4} + \cdots + \frac{x_{n-2} x_{n-1}}{x_n} + \frac{x_{n-1}}{x_2}(x_3 + x_4 + \cdots + x_n). \end{aligned}$$

Hence,

$$x_1 + x_2 + \cdots + x_{n-2} + x_{n-1} \leqslant \frac{x_1 x_2}{x_3} + \cdots + \frac{x_{n-2} x_{n-1}}{x_n} + \frac{x_{n-1} x_n}{x_2}. \quad ③$$

On the other hand, let $(x, y, a) = \left(x_n, x_n \cdot \frac{x_{n-1}}{x_2}, \frac{x_1}{x_2}\right)$, we have

$$x_n + \frac{x_n x_{n-1}}{x_2} \leqslant \frac{x_n x_1}{x_2} + \frac{x_{n-1} x_n}{x_1}. \quad ④$$

Adding ③ and ④, the original inequality is proved.

Eg. 9 Assume n is a positive integer greater than 3. $a_1, a_2, \cdots, a_n$ are real numbers satisfying $2 \leqslant a_i \leqslant 3$, $i = 1, 2, \cdots, n$. Denote $S = a_1 + a_2 + \cdots + a_n$. Prove that

$$\frac{a_1^2 + a_2^2 - a_3^2}{a_1 + a_2 - a_3} + \frac{a_2^2 + a_3^2 - a_4^2}{a_2 + a_3 - a_4} + \cdots + \frac{a_n^2 + a_1^2 - a_2^2}{a_n + a_1 - a_2} \leqslant 2S - 2n.$$

Hint. To handle the $2S$ part on the right hand side of the inequality, we can separate $a_i + a_{i+1} + a_{i+2}$ from each term on the left hand side, whose partial sum adds up to $2S$.

Proof.

$$\frac{a_i^2 + a_{i+1}^2 - a_{i+2}^2}{a_i + a_{i+1} - a_{i+2}} = a_i + a_{i+1} + a_{i+2} - \frac{2a_i a_{i+1}}{a_i + a_{i+1} - a_{i+2}}.$$

Note that

$$1 = 2 + 2 - 3 \leqslant a_i + a_{i+1} + a_{i+2} \leqslant 3 + 3 - 2 = 4,$$

along with $(a_i - 2)(a_{i+1} - 2) \geqslant 0$, we obtain

$$-2a_i a_{i+1} \leqslant -4(a_i + a_{i+1} - 2).$$

Thus,

$$\begin{aligned}
&\frac{a_i^2 + a_{i+1}^2 - a_{i+2}^2}{a_i + a_{i+1} - a_{i+2}} \\
&\leqslant a_i + a_{i+1} + a_{i+2} - 4 \cdot \frac{a_i + a_{i+1} - 2}{a_i + a_{i+1} - a_{i+2}} \\
&= a_i + a_{i+1} + a_{i+2} - 4\left(1 + \frac{a_{i+2} - 2}{a_i + a_{i+1} - a_{i+2}}\right)
\end{aligned}$$

$$\leqslant a_i + a_{i+1} + a_{i+2} - 4\left(1 + \frac{a_{i+2} - 2}{4}\right)$$

$$= a_i + a_{i+1} - 2.$$

Denote $a_{n+1} = a_1$, $a_{n+2} = a_2$. Replace i by $1, 2, \cdots, n$ in the above inequality and we get n inequalities. Now add them up, and we have

$$\frac{a_1^2 + a_2^2 - a_3^2}{a_1 + a_2 - a_3} + \frac{a_2^2 + a_3^2 - a_4^2}{a_2 + a_3 - a_4} + \cdots + \frac{a_n^2 + a_1^2 - a_2^2}{a_n + a_1 - a_2} \leqslant 2S - 2n.$$

Exercise 6

1. Let $x_1, x_2, \cdots, x_n$ be positive real numbers. Prove that

$$\frac{1}{x_1} + \frac{1}{x_2} + \cdots + \frac{1}{x_n}$$

$$\geqslant 2\left(\frac{1}{x_1 + x_2} + \frac{1}{x_2 + x_3} + \cdots + \frac{1}{x_{n-1} + x_n} + \frac{1}{x_n + x_1}\right).$$

2. Let a, b, c be positive real numbers, prove that

$$\frac{1}{a^3 + b^3 + abc} + \frac{1}{b^3 + c^3 + abc} + \frac{1}{c^3 + a^3 + abc} \leqslant \frac{1}{abc}.$$

3. Given that $0 \leqslant x, y, z \leqslant 1$, solve the following equation:

$$\frac{x}{1 + y + zx} + \frac{y}{1 + z + xy} + \frac{z}{1 + x + yz} = \frac{3}{x + y + z}.$$

4. Let a, b, c be positive real numbers and $abc = 1$. Prove that

$$\sum \frac{ab}{a^5 + b^5 + ab} \leqslant 1.$$

When does the equality hold?

5. Given $\alpha, \beta > 0$, $x, y, z \in \mathbf{R}^+$, $xyz = 2004$. Find the maximum for u, which is defined as:

$$u = \sum \frac{1}{2004^{\alpha+\beta} + x^{\alpha}(y^{2\alpha+3\beta} + z^{2\alpha+3\beta})}.$$

6. Let $x_1, x_2, \cdots, x_n$ be positive numbers such that $x_1 + x_2 + \cdots + x_n = a$. For $m, n \in \mathbf{Z}^+$, $m, n > 1$, prove that

$$\frac{x_1^m}{a - x_1} + \frac{x_2^m}{a - x_2} + \cdots + \frac{x_n^m}{a - x_n} \geqslant \frac{a^{m-1}}{(n-1)n^{m-2}}.$$

7. Let a, b, c be positive real numbers, prove that

(1) $\sqrt[3]{\dfrac{a}{b+c}} + \sqrt[3]{\dfrac{b}{c+a}} + \sqrt[3]{\dfrac{c}{a+b}} > \dfrac{3}{2}$.

(2) $\sqrt[3]{\dfrac{a^2}{(b+c)^2}} + \sqrt[3]{\dfrac{b^2}{(c+a)^2}} + \sqrt[3]{\dfrac{c^2}{(a+b)^2}} \geqslant \dfrac{3}{\sqrt[3]{4}}$.

8. Assume integer $n \geqslant 2$. Given n positive numbers $v_1, v_2, \cdots, v_n$ satisfying the following two conditions:

(1) $v_1 + v_2 + \cdots + v_n = 1$.

(2) $v_1 \leqslant v_2 \leqslant \cdots \leqslant v_n \leqslant 2v_1$.

Find the maximum for $v_1^2 + v_2^2 + \cdots + v_n^2$.

9. Assume $a > 1$, $b > 1$, $c > 1$, show that

(1) $\dfrac{a^5}{b^2 - 1} + \dfrac{b^5}{c^2 - 1} + \dfrac{c^5}{a^2 - 1} \geqslant \dfrac{25}{6}\sqrt{15}$.

(2) $\dfrac{a^5}{b^3 - 1} + \dfrac{b^5}{c^3 - 1} + \dfrac{c^5}{a^3 - 1} \geqslant \dfrac{5}{2}\sqrt[3]{50}$.

10. (The Reverse Cauchy-Schwarz Inequality: The Polya-Szego Inequality) Assume $0 < m_1 \leqslant a_i \leqslant M_1$, $0 < m_2 \leqslant b_i \leqslant M_2$, $i = 1, 2, \cdots, n$. Then,

$$\frac{\left(\sum_{i=1}^{n} a_i^2\right)\left(\sum_{i=1}^{n} b_i^2\right)}{\left(\sum_{i=1}^{n} a_i b_i\right)^2} \leqslant \frac{1}{4} \cdot \left(\sqrt{\frac{M_1 M_2}{m_1 m_2}} + \sqrt{\frac{m_1 m_2}{M_1 M_2}}\right)^2.$$

11. Let $x_1, x_2, \cdots, x_n$ be positive real numbers satisfying $\sum_{i=1}^{n} x_i = 1$. Prove that

$$\sum_{i=1}^{n} \sqrt{\frac{1}{x_i} - 1} \geqslant (n-1) \cdot \sum_{i=1}^{n} \frac{1}{\sqrt{\dfrac{1}{x_i} - 1}}.$$

12. Assume all the elements in the finite sets S_1, S_2, $\cdots$, S_n are non-negative integers. x_i stands for the sum of all elements in S_i. Prove that if for some integer k, $1 < k < n$, we have

$$\sum_{i=1}^{n} x_i \leqslant \frac{1}{k+1} \cdot \left[k \frac{n(n+1)(2n+1)}{6} - (k+1)^2 \frac{n(n+1)}{2}\right],$$

then there exist subscripts i, j, t, l (at least three of them are mutually different), such that

$$x_i + x_j = x_t + x_l.$$

Chapter 7 Mathematical Induction and Inequality

There are several variants of the Mathematical Induction. The most elementary and useful ones are the First and the Second Principle of Mathematical Induction.

The First Principle of Mathematical Induction: Assume $P(n)$ is a statement (with respect to the positive integer n). If (1) $P(1)$ holds; (2) Assume $P(k)$ holds, we can show that $P(k+1)$ holds, then $P(n)$ holds for any positive integer n.

The Second Principle of Mathematical Induction: Assume $P(n)$ is a statement (with respect to the positive integer n). If (1) $P(1)$ holds; (2) Assume for $n \leqslant k$ (k is any positive integer) $P(n)$ holds, and we can imply that $P(k+1)$ holds, then $P(n)$ holds for any positive integer n.

When dealing with an inequality related with positive integer n, we can always consider using the mathematical induction.

Eg. 1 Assume $a_1 = 2$, $a_{n+1} = \frac{a_n}{2} + \frac{1}{a_n}$ $(n = 1, 2, \cdots, 2004)$. Prove that

$$\sqrt{2} < a_{2005} < \sqrt{2} + \frac{1}{2005}.$$

Proof. According to the AM-GM Inequality, we have

$$a_{2005} = \frac{a_{2004}}{2} + \frac{1}{a_{2004}} > 2\sqrt{\frac{a_{2004}}{2} \cdot \frac{1}{a_{2004}}} = \sqrt{2}.$$

Next we prove using induction that, for any positive integer n,

$$\sqrt{2} < a_n < \sqrt{2} + \frac{1}{n}. \tag{①}$$

The base case when $n = 1$ is obvious.

Assume when $n = k$, we have $\sqrt{2} < a_k < \sqrt{2} + \frac{1}{k}$. Then, when $n = k + 1$,

$$a_{k+1} = \frac{a_k}{2} + \frac{1}{a_k} < \frac{\sqrt{2} + \frac{1}{k}}{2} + \frac{1}{\sqrt{2}} = \sqrt{2} + \frac{1}{2k} \leqslant \sqrt{2} + \frac{1}{k+1}.$$

Thus ① holds for any positive integer n, in our case n can be 2005.

Eg. 2 Assume $a > 0$. Prove that, for any positive integer n, we have the inequality

$$\frac{1 + a^2 + \cdots + a^{2n}}{a + a^3 + \cdots + a^{2n-1}} \geqslant \frac{n+1}{n}.$$

Proof. Apply mathematical induction. When $n = 1$, $\frac{1 + a^2}{a} \geqslant 2$, so the inequality holds.

Assume that when $n = k$, the inequality holds, or

$$A = \frac{1 + a^2 + \cdots + a^{2k}}{a + a^3 + \cdots + a^{2k-1}} > \frac{k+1}{k}.$$

Our goal is to prove, when $n = k + 1$,

$$B = \frac{1 + a^2 + \cdots + a^{2k+2}}{a + a^3 + \cdots + a^{2k+1}} > \frac{k+2}{k+1}.$$

Unfortunately A and B have different denominators, how should we relate them then? A natural idea is to look at the reciprocal of A. Note that

$$\frac{1}{A} < \frac{k}{k+1}.$$

However,

$$\begin{aligned}\frac{1}{A} + B &= \frac{a + a^3 + \cdots + a^{2k-1}}{1 + a^2 + \cdots + a^{2k}} + \frac{1 + a^2 + \cdots + a^{2k+2}}{a + a^3 + \cdots + a^{2k+1}} \\ &= \frac{1 + 2a^2 + 2a^4 + \cdots + 2a^{2k} + a^{2k+2}}{a + a^3 + \cdots + a^{2k+1}}\end{aligned}$$

$$= \frac{(1+a^2)+(a^2+a^4)+\cdots+(a^{2k-2}+a^{2k})+(a^{2k}+a^{2k+2})}{a+a^3+\cdots+a^{2k+1}}$$

$\geqslant 2.$

Hence,

$$B > 2 - \frac{1}{A} > 2 - \frac{k}{k+1} = \frac{k+2}{k+1}.$$

So the inequality holds for $n = k+1$. Therefore, the original inequality holds for any positive integer n.

Eg. 3 Assume a sequence $\{a_n\}$ is composed of positive terms which satisfy $a_n^2 \leqslant a_n - a_{n+1}$ for any $n \in \mathbf{N}_+$. Prove that $a_n < \frac{1}{n}$.

Proof. First, since $a_{n+1} \leqslant a_n - a_n^2 = a_n(1-a_n)$ and $\{a_n\}$ is composed of positive terms, we have

$$a_n(1-a_n) > 0 \Rightarrow 0 < a_n < 1.$$

Next, applying mathematical induction, we show that $a_n < \frac{1}{n}$.

When $n = 1$, the above statement is obviously true.

When $n = 2$, if $\frac{1}{2} \leqslant a_1 < 1$, then $0 < 1 - a_1 \leqslant \frac{1}{2}$. If $0 < a_1 < \frac{1}{2}$, then $\frac{1}{2} < 1 - a_1 < 1$. Whenever the case, we always have

$$a_2 \leqslant a_1(1-a_1) < \frac{1}{2}.$$

Hence the statement is also true.

Now, assume that when $n = k$, $a_k < \frac{1}{k}$ $(k \geqslant 2)$ holds.

Thus, when $n = k+1$,

$$a_{k+1} \leqslant a_k - a_k^2 = \frac{1}{4} - \left(a_k - \frac{1}{2}\right)^2.$$

Since $0 < a_k < \frac{1}{k}$, we have

$$-\frac{1}{2} < a_k - \frac{1}{2} < \frac{1}{k} - \frac{1}{2} \leqslant 0.$$

Hence,

$$0 < \frac{1}{4} - \left(a_k - \frac{1}{2}\right)^2 < -\frac{1-k}{k^2} < \frac{k-1}{k^2-1} = \frac{1}{k+1}.$$

Therefore $a_{k+1} < \frac{1}{k+1}$, the statement is also true for $k+1$.

In sum, for any $n \in \mathbf{N}_+$, we have $a_n < \frac{1}{n}$.

Note. We can use a different approach in the inductive step (i. e. from k to $k+1$). And later we provide a proof without using the induction.

[2nd Proof] First it's easy to see that $0 < a_n < 1$. The base case is the same as we have shown before.

Assume that when $n = k$, we have $a_k < \frac{1}{k}$.

Now divide the interval $\left(0, \frac{1}{k}\right)$ into two sub-intervals $\left(0, \frac{1}{k+1}\right)$ and $\left[\frac{1}{k+1}, \frac{1}{k}\right)$. Hence, when $n = k+1$,

(i) On $\left(0, \frac{1}{k+1}\right)$, $a_{k+1} \leqslant a_k(1-a_k) < a_k < \frac{1}{k+1}$.

(ii) On $\left[\frac{1}{k+1}, \frac{1}{k}\right)$, $a_{k+1} \leqslant a_k(1-a_k) < \frac{1}{k}\left(1 - \frac{1}{k+1}\right) = \frac{1}{k+1}$.

Combining these two, we have for any n, $a_n < \frac{1}{n}$.

[3rd Proof] Since $a_n^2 \leqslant a_n - a_{n-1} \Rightarrow a_{n+1} \leqslant a_n(1-a_n)$. Thus $0 < a_n < 1$, and $a_n > a_{n+1}$.

As a result,

$$\frac{1}{a_{n+1}} \geqslant \frac{1}{a_n(1-a_n)} = \frac{1}{1-a_n} + \frac{1}{a_n},$$

or

$$\frac{1}{a_{n+1}} - \frac{1}{a_n} = \frac{1}{1-a_n} > 1.$$

So

$$\frac{1}{a_n}-\frac{1}{a_1}=\sum_{k=1}^{n}\left(\frac{1}{a_{k+1}}-\frac{1}{a_k}\right)>n-1.$$

Therefore,

$$\frac{1}{a_n}>(n-1)+\frac{1}{a_1}>(n-1)+1=n\Rightarrow a_n<\frac{1}{n}.$$

Eg. 4 Assume a non-negative sequence a_1, a_2, $\cdots$ satisfies the following condition: $a_{m+n}\leqslant a_n+a_m$, m, $n\in\mathbf{N}_+$. Prove that, for any positive integer n, we have

$$a_n\leqslant ma_1+\left(\frac{n}{m}-1\right)a_m. \quad ⊛$$

Proof. Let $m=1$, we have

$$0\leqslant a_n\leqslant a_{n-1}+a_1\leqslant\cdots\leqslant na_1. \quad ①$$

Next, applying mathematical induction, we prove that the statement is true for any n, $m\in\mathbf{N}_+$.

Fix $m\in\mathbf{N}_+$, pick $n=1$, we need to prove that

$$\left(1-\frac{1}{m}\right)a_m\leqslant(m-1)a_1,$$

or

$$(m-1)a_m\leqslant(m-1)ma_1.$$

From ① we know the above is true, hence the statement is true for $n=1$.

Now assume the statement is true for $1\leqslant n\leqslant k$. We discuss two situations.

(i) $k<m$, from ⊛ we know

$$\frac{a_{k+1}}{k+1}-\frac{a_m}{m}\leqslant a_1-\frac{a_m}{m}.$$

But $a_1-\frac{a_m}{m}\geqslant 0$, we have

$$\frac{a_{k+1}}{k+1} - \frac{a_m}{m} \leqslant \frac{m}{k+1}\left(a_1 - \frac{a_m}{m}\right).$$

Thus,

$$a_{k+1} \leqslant ma_1 + \left(\frac{k+1}{m} - 1\right)a_m.$$

(ii) $k \geqslant m$, then $k+1-m \geqslant 1$, from the given conditions, we have

$$a_{k+1} \leqslant a_{k+1-m} + a_m.$$

Using the inductive hypothesis,

$$a_{k+1-m} \leqslant ma_1 + \left(\frac{k+1-m}{m} - 1\right)a_m.$$

Therefore,

$$a_{k+1} \leqslant ma_1 + \left(\frac{k+1}{m} - 1\right)a_m.$$

Combining (i) and (ii), we see that the statement holds for $n = k+1$.

Note. We can also prove the original inequality using the Division Algorithm.

[2nd Proof] From the given condition,

$$0 \leqslant a_n \leqslant a_{n-1} + a_1 \leqslant \cdots \leqslant na_1.$$

Let $n = mq + r$, then

$$\begin{aligned} ma_n &= ma_{mq+r} \\ &\leqslant ma_{mq} + ma_r \\ &\leqslant mqa_m + ma_r \\ &= (n-r)a_m + ma_r \\ &= (n-m)a_m + (m-r)a_m + ma_r \\ &\leqslant (n-m)a_m + (m-r)ma_1 + mra_1 \\ &= (n-m)a_m + m^2a_1. \end{aligned}$$

Hence,

$$a_n \leqslant ma_1 + \left(\frac{n}{m} - 1\right)a_m.$$

Eg. 5 Prove that for any positive integer n, we have

$$\sqrt{1^2 + \sqrt{2^2 + \sqrt{3^2 + \cdots + \sqrt{n^2}}}} < 2.$$

Hint. It's hard to handle this problem directly. Instead, we apply the backward induction and start with the most inner square root.

Proof. Let us prove that

$$\sqrt{k^2 + \sqrt{(k+1)^2 + \cdots + \sqrt{n^2}}} < k + 1. \quad ①$$

When $k = n$, ① obviously holds.

Now, assume ① holds for $k = n, n-1, \cdots, m$, thus when $k = m - 1$, it suffices to show that

$$\sqrt{(m-1)^2 + \sqrt{m^2 + \cdots + \sqrt{n^2}}} < m - 1 + 1 = m.$$

According to the inductive hypothesis, the left hand side of the above formula $< \sqrt{(m-1)^2 + m + 1} = \sqrt{m^2 - m + 2} \leqslant m$ $(m \geqslant 2)$.

Hence when $k = m - 1$, ① is true.

As a result, ① should also hold for $k = 1$, so the original inequality is proved.

Note. Backward Induction means: assume $P(n)$ is a statement related with positive integer n. If:

(1) $P(n)$ holds for infinitely many integers n,

(2) Assume $P(k+1)$ holds, we can show that $P(k)$ holds.

Then $P(n)$ holds for any positive integer n.

Eg. 6 Assume $\{a_k\}(k \geqslant 1)$ is a sequence of positive real numbers, and there exists a constant k, such that $a_1^2 + a_2^2 + \cdots + a_n^2 < ka_{n+1}^2$ (for all $n \geqslant 1$). Prove that there exists a constant c, such that $a_1 + a_2 + \cdots + a_n < ca_{n+1}$ (for all $n \geqslant 1$).

Proof. Consider a chain of inequalities:

$$(a_1 + a_2 + \cdots + a_n)^2 < t(a_1^2 + a_2^2 + \cdots + a_n^2) < c^2 a_{n+1}^2.$$

We know that

$$t(a_1^2 + a_2^2 + \cdots + a_n^2) < tk \cdot a_{n+1}^2,$$

so we only need to make $tk = c^2$, where t is a parameter.

Assume statement P_i as the following:

$$(a_1 + a_2 + \cdots + a_i)^2 < t(a_1^2 + a_2^2 + \cdots + a_i^2),$$

and define statement Q_i as:

$$a_1 + a_2 + \cdots + a_i < ca_{i+1}.$$

When $i = 1$, to make P_1 true, we require $t > 1$.

Assume now that statement P_k holds, or

$$(a_1 + a_2 + \cdots + a_k)^2 < t(a_1^2 + a_2^2 + \cdots + a_k^2).$$

From the chain of inequalities, we have

$$(a_1 + a_2 + \cdots + a_k)^2 < tka_{k+1}^2 = c^2 a_{k+1}^2.$$

Thus,

$$a_1 + a_2 + \cdots + a_k < ca_{k+1}.$$

What we just show is that P_k holds $\Rightarrow Q_k$ holds.

Naturally, we want to show that if Q_k holds $\Rightarrow P_{k+1}$ holds.

In other words, we want to show

$$a_1 + a_2 + \cdots + a_k < ca_{k+1}$$
$$\Rightarrow (a_1 + a_2 + \cdots + a_{k+1})^2 < t(a_1^2 + a_2^2 + \cdots + a_{k+1}^2).$$

For the above inequality to hold, it suffices to have

$$a_{k+1}^2 + 2a_{k+1}(a_1 + a_2 + \cdots + a_k) < ta_{k+1}^2,$$

or

$$a_1 + a_2 + \cdots + a_k < \frac{t-1}{2} a_{k+1}.$$

As a result, we can pick $c = \dfrac{t-1}{2}$.

Now,

$$c = \frac{t-1}{2} = \frac{\frac{c^2}{k} - 1}{2} \Rightarrow c^2 - 2kc - k = 0.$$

Choose $c = k + \sqrt{k^2 + k}$, then $t = \frac{c^2}{k} > 1$ satisfies the requirement, therefore by induction we have proved the conclusion.

Note. Here we have used another form of the induction called the Spiral Induction:

Assume that $P(n)$, $Q(n)$ are two sets of statements. If:

(1) Statement $P(1)$ holds.

(2) For any positive integer k, statement $P(k)$ holds implies statement $Q(k)$ holds. Also, statement $Q(k)$ holds implies statement $P(k+1)$ holds.

It follows that for any positive integer n, statements $P(n)$ and $Q(n)$ both hold true.

Eg. 7 Assume $a_1 < a_2 < \cdots < a_n$ are real numbers. Prove that:

$$a_1 a_2^4 + a_2 a_3^4 + \cdots + a_n a_1^4 \geqslant a_2 a_1^4 + a_3 a_2^4 + \cdots + a_1 a_n^4.$$

Proof. Apply mathematical induction.

When $n = 2$, the equal sign holds.

Now assume the inequality holds for $n-1$, or

$$a_1 a_2^4 + a_2 a_3^4 + \cdots + a_{n-1} a_1^4 \geqslant a_2 a_1^4 + a_3 a_2^4 + \cdots + a_1 a_{n-1}^4.$$

Consider the case for n. It suffices to show that

$$a_{n-1} a_n^4 + a_n a_1^4 - a_{n-1} a_1^4 \geqslant a_n a_{n-1}^4 + a_1 a_n^4 - a_1 a_{n-1}^4. \quad ①$$

(Note that this is exactly the case when $n = 3$!)

WLOG, assume $a_n - a_1 = 1$, otherwise we can divide a_1, a_{n-1}, a_n each by $a_n - a_1 > 0$, which will not affect the inequality.

Using Jensen's Inequality on the convex function x^4, we obtain:

$$a_1^4(a_n - a_{n-1}) + a_n^4(a_{n-1} - a_1) \geqslant (a_1(a_n - a_{n-1}) + a_n(a_{n-1} - a_1))^4.$$

Simplifying the above inequality, it's easy to see that ① holds.

Eg. 8 Define a sequence $x_1, x_2, \cdots, x_n$ as follows: $x_1 \in [0, 1)$, and

$$x_{n+1} = \begin{cases} \frac{1}{x_n} - \left[\frac{1}{x_n}\right], & \text{if } x_n \neq 0, \\ 0, & \text{if } x_n = 0. \end{cases}$$

Prove that, for any positive integer n, we have

$$x_1 + x_2 + \cdots + x_n < \frac{f_1}{f_2} + \frac{f_2}{f_3} + \cdots + \frac{f_n}{f_{n+1}},$$

where $\{f_n\}$ is the Fibonacci sequence defined as: $f_1 = f_2 = 1$, $f_{n+2} = f_{n+1} + f_n$, $n \in \mathbf{N}_+$.

Proof. When $n = 1$, since $x_1 \in [0, 1)$, we have $x_1 < 1 = \frac{f_1}{f_2}$, the statement is true.

When $n = 2$, we consider the following two cases:

(1) If $x_1 \leqslant \frac{1}{2}$, then $x_1 + x_2 < \frac{1}{2} + 1 = \frac{3}{2} = \frac{f_1}{f_2} + \frac{f_2}{f_3}$.

(2) If $\frac{1}{2} < x_1 < 1$, then $x_1 + x_2 = x_1 + \frac{1}{x_1} - 1$.

Let $f(t) = t + \frac{1}{t}$, then $f(t)$ is monotonically decreasing on $\left[\frac{1}{2}, 1\right)$. Thus,

$$x_1 + x_2 = f(x_1) - 1 < f\left(\frac{1}{2}\right) - 1 = \frac{3}{2}.$$

Hence the statement is also true for $n = 2$.

Assume when $n = k$ and $n = k + 1$, the statement is true. So we have

$$x_1 + x_2 + \cdots + x_k < \frac{f_1}{f_2} + \frac{f_2}{f_3} + \cdots + \frac{f_k}{f_{k+1}},$$

$$x_1 + x_2 + \cdots + x_{k+1} < \frac{f_1}{f_2} + \frac{f_2}{f_3} + \cdots + \frac{f_{k+1}}{f_{k+2}}.$$

Replace x_1 by x_2, x_2 by x_3, so on and so forth, we have

$$x_2+x_3+\cdots+x_{k+2}<\frac{f_1}{f_2}+\frac{f_2}{f_3}+\cdots+\frac{f_{k+1}}{f_{k+2}}. \qquad ①$$

Similarly,

$$x_3+x_4+\cdots+x_{k+2}<\frac{f_1}{f_2}+\frac{f_2}{f_3}+\cdots+\frac{f_k}{f_{k+1}}. \qquad ②$$

Again, we discuss two possible situations:

(i) If $x_1\leqslant\frac{f_{k+2}}{f_{k+3}}$, from ① we obtain:

$$x_1+x_2+\cdots+x_{k+2}<\frac{f_1}{f_2}+\frac{f_2}{f_3}+\cdots+\frac{f_{k+2}}{f_{k+3}}.$$

So the statement is true for $n=k+2$.

(ii) If $x_1>\frac{f_{k+2}}{f_{k+3}}$, then $x_1\in\left(\frac{f_{k+2}}{f_{k+3}},1\right)$. Hence

$$x_1+x_2=x_1+\frac{1}{x_1}-1<\frac{f_{k+2}}{f_{k+3}}+\frac{f_{k+3}}{f_{k+2}}-1=\frac{f_{k+2}}{f_{k+3}}+\frac{f_{k+1}}{f_{k+2}}.$$

From ②, we also have

$$x_1+x_2+\cdots+x_{k+2}<\frac{f_1}{f_2}+\frac{f_2}{f_3}+\cdots+\frac{f_{k+2}}{f_{k+3}}.$$

Therefore the statement is true for $n=k+2$.

Note. We can also use a more complicated method involving continued fraction, as shown in the following proof:

[2nd Proof] Denote $f(x)=\frac{1}{1+x}$. Let

$$g_n(x)=x+f(x)+f^{(2)}(x)+\cdots+f^{(n)}(x),\ n=0,1,2,\cdots.$$

Note that $\frac{f_1}{f_2}=1$ and $f\left(\frac{f_i}{f_{i+1}}\right)=\frac{1}{1+\frac{f_i}{f_{i+1}}}=\frac{f_{i+1}}{f_{i+2}}$, we obtain

$$f^{(k)}(1)=f^{(k)}\left(\frac{f_1}{f_2}\right)=f^{(k-1)}\left(\frac{f_2}{f_3}\right)=\cdots=f\left(\frac{f_k}{f_{k+1}}\right)=\frac{f_{k+1}}{f_{k+2}},$$

$$k=1,2,\cdots.$$

Thus,

$$g_{n-1}(1)=\frac{f_1}{f_2}+\frac{f_2}{f_3}+\cdots+\frac{f_n}{f_{n+1}}.$$

Next we prove a lemma:

Lemma (I) For any $x, y \in [0, 1]$, if $x \neq y$, then $|f(x)-f(y)|<|x-y|$, and $f(x)-f(y)$ has a different sign than $x-y$.

Lemma (II) $g_n(x)$ is monotonically increasing on $[0, 1]$.

Proof of Lemma:

(I) since $f(x)-f(y)=\dfrac{y-x}{(1+x)(1+y)}$, this is easy to verify.

(II) If $x>y$, from (I) we know that in the expression

$$g_n(x)-g_n(y)=(x-y)+(f(x)-f(y))+\cdots+(f^{(n)}(x)-f^{(n)}(y)).$$

Note that the absolute value of each subtraction term is less than the previous one, and with opposite sign. Since $x-y>0$, we conclude that $g_n(x)-g_n(y)>0$. Back to the original problem, if each $x_i(i<n)=0$, then $x_n=0$. From the inductive hypothesis of the previous $n-1$ terms, it's easy to verify the conclusion. Otherwise for $2\leqslant i\leqslant n$, we have:

$$x_{i-1}=\frac{1}{a_i+x_i},$$

where $a_i=\left[\dfrac{1}{x_{i-1}}\right]$ is an integer.

Thus,

$$\begin{aligned}&x_n+x_{n-1}+x_{n-2}+\cdots+x_1\\&=x_n+\frac{1}{a_n+x_n}+\frac{1}{a_{n-1}+\frac{1}{a_n+x_n}}+\cdots+\frac{1}{a_2+\frac{1}{a_3+\frac{1}{\cdots+\frac{1}{a_n+x_n}}}}\end{aligned}$$

$\triangleq S.$

We use induction to prove that, if fix $x_n \in [0, 1)$, S reaches its maximum when $a_i = 1$ for all i.

First, a_2 only appears in the final term of S and it does not interact with x_n, a_n, $\cdots$, a_3, so S is maximized when $a_2 = 1$.

Now assume that for $i > 2$, S is maximized with $a_{i-1} = a_{i-2} = \cdots = a_2 = 1$, and such maximum does not depend on the value of x_n, a_n, a_{n-1}, $\cdots$, a_i.

We note that only the later $i - 1$ terms in S contains a_i, and these terms add up to

$$g_{i-2}\left(\cfrac{1}{a_i + \cfrac{1}{a_{i+1} + \cfrac{1}{\cdots + \cfrac{1}{a_n + x_n}}}}\right).$$

Since g_{i-2} is an increasing function, it will be maximized when a_i is minimized, or when $a_i = 1$.

Finally, from what we have discussed above, we have

$$\begin{aligned}
& x_n + x_{n-1} + \cdots + x_1 \\
\leqslant & x_n + \frac{1}{1 + x_n} + \cfrac{1}{1 + \cfrac{1}{1 + x_n}} + \cdots + \cfrac{1}{1 + \cfrac{1}{1 + \cfrac{1}{\cdots + \cfrac{1}{1 + x_n}}}} \\
= & g_{n-1}(x_n) < g_{n-1}(1) \\
= & \frac{f_1}{f_2} + \frac{f_2}{f_3} + \cdots + \frac{f_n}{f_{n+1}}.
\end{aligned}$$

Eg. 9 Given a sequence $\{r_n\}$ that satisfies: $r_1 = 2$, $r_n = r_1 r_2 \cdots r_{n-1} + 1 (n = 2, 3, \cdots)$.

Also, positive integers a_1, a_2, $\cdots$, a_n satisfy $\sum_{k=1}^{n} \frac{1}{a_k} < 1$. Prove that

$$\sum_{i=1}^{n} \frac{1}{a_i} \leqslant \sum_{i=1}^{n} \frac{1}{r_i}.$$

Proof. From the definition of $\{r_n\}$, it's easy to see that

$$1-\frac{1}{r_1}-\frac{1}{r_2}-\cdots-\frac{1}{r_n}=\frac{1}{r_1r_2\cdots r_n}. \qquad ①$$

When $\frac{1}{a_1}<1$, positive integer $a_1 \geqslant 2 = r_1$, thus $\frac{1}{a_1} \leqslant \frac{1}{r_1}$.

Assume when $n<k$, the original inequality holds for any positive integers a_1, a_2, $\cdots$, a_n that satisfy the condition.

When $n=k$, if there exists a sequence of positive integers (that satisfy the condition) a_1, a_2, $\cdots$, a_n, such that

$$\frac{1}{a_1}+\frac{1}{a_2}+\cdots+\frac{1}{a_n}>\frac{1}{r_1}+\frac{1}{r_2}+\cdots+\frac{1}{r_n}. \qquad ②$$

WLOG, assume $a_1 \leqslant a_2 \leqslant \cdots \leqslant a_n$.

Then, according to the inductive hypothesis,

$$\frac{1}{a_1} \leqslant \frac{1}{r_1},$$

$$\frac{1}{a_1}+\frac{1}{a_2} \leqslant \frac{1}{r_1}+\frac{1}{r_2},$$

$$\cdots\cdots$$

$$\frac{1}{a_1}+\frac{1}{a_2}+\cdots+\frac{1}{a_{n-1}} \leqslant \frac{1}{r_1}+\frac{1}{r_2}+\cdots+\frac{1}{r_{n-1}}.$$

Multiply the above inequalities by non-negative numbers a_1-a_2, a_2-a_3, $\cdots$, $a_{n-1}-a_n$ alternatively, and multiply ② by a_n. Add the results up, we obtain

$$n>\frac{a_1}{r_1}+\frac{a_2}{r_2}+\cdots+\frac{a_n}{r_n}.$$

Hence,

$$1>\frac{1}{n}\left(\frac{a_1}{r_1}+\cdots+\frac{a_n}{r_n}\right)$$

$$\geqslant \sqrt[n]{\frac{a_1a_2\cdots a_n}{r_1r_2\cdots r_n}},$$

or

$$r_1 r_2 \cdots r_n \geqslant a_1 a_2 \cdots a_n. \qquad ③$$

On the other hand, the positive number $1-\left(\frac{1}{a_1}+\frac{1}{a_2}+\cdots+\frac{1}{a_n}\right) \geqslant \frac{1}{a_1 a_2 \cdots a_n}$, so combining ① and ②, we have:

$$\frac{1}{r_1 r_2 \cdots r_n} \geqslant \frac{1}{a_1 a_2 \cdots a_n}. \qquad ④$$

③ and ④ clearly lead to a contradiction!

Note. Reader can check that the equality holds if and only if $\{a_1, a_2, \cdots, a_n\} = \{r_1, r_2, \cdots, r_n\}$.

This problem is a conjuncture by Erdös. The above solution includes a proof by contradiction in the inductive steps, which is equivalent to adding an extra condition that will help us in proving the inequality.

From what we have proved in Eg. 9, we can also verify that the following two statements are true:

(A) Assume $x_1 \geqslant x_2 \geqslant \cdots \geqslant x_n > 0$, $y_1 \geqslant y_2 \geqslant \cdots \geqslant y_n > 0$, and $\sum_{i=1}^{n} x_i \leqslant \sum_{i=1}^{n} y_i$, $\sum_{i=1}^{k} x_i \geqslant \sum_{i=1}^{k} y_i \, (k = 1, 2, \cdots, n-1)$. Then,

$$\prod_{i=1}^{n} x_i \leqslant \prod_{i=1}^{n} y_i.$$

(B) Assume $a_1 \geqslant a_2 \geqslant \cdots \geqslant a_n > 0$, and $\prod_{i=1}^{k} b_i \geqslant \prod_{i=1}^{k} a_i \, (1 \leqslant k \leqslant n)$. Then,

$$\sum_{i=1}^{n} b_i \geqslant \sum_{i=1}^{n} a_i.$$

Exercise 7

1. Assume $a_1 = 3$, $a_n = a_{n-1}^2 - n \, (n = 2, 3, \cdots)$. Prove that $a_n > 0$.
2. Assume a sequence of numbers $a_1, a_2, \cdots, a_{2n+1}$ satisfy $a_i -$

$2a_{i+1} + a_{i+2} \geqslant 0$ ($i = 1, 2, \cdots, 2n-1$). Prove that

$$\frac{a_1 + a_3 + \cdots + a_{2n+1}}{n+1} \geqslant \frac{a_2 + a_4 + \cdots + a_{2n}}{n}.$$

3. Assume $\{a_n\}$ is a sequence of positive numbers. If $a_{n+1} \leqslant a_n - a_n^2$, show that for any $n \geqslant 2$, we have

$$a_n \leqslant \frac{1}{n+2}.$$

4. Given a sequence $\{a_n\}$, $a_1 = a_2 = 1$, $a_{n+2} = a_{n+1} + a_n$. Prove that, for any $n \in \mathbf{N}_+$, we have

$$\operatorname{arccot} a_n \leqslant \operatorname{arccot} a_{n+1} + \operatorname{arccot} a_{n+2}.$$

Point out when does the equality hold?

5. Assume $a_1, a_2, \cdots$ is a real sequence. For any $i, j = 1, 2, \cdots$, we have $a_{i+j} \leqslant a_i + a_j$. Prove that for positive integer n,

$$a_1 + \frac{a_2}{2} + \frac{a_3}{3} + \cdots + \frac{a_n}{n} \geqslant a_n.$$

6. Assume non-negative integers $a_1, a_2, \cdots, a_{2004}$ satisfy $a_i + a_j \leqslant a_{i+j} \leqslant a_i + a_j + 1$ ($1 \leqslant i, j, i+j \leqslant 2004$). Prove that there exists $x \in \mathbf{R}$, such that for any n ($1 \leqslant n \leqslant 2004$), we have $a_n = [nx]$.

7. Assume $1 < x_1 < 2$. For $n = 1, 2, 3, \cdots$, define $x_{n+1} = 1 + x_n - \frac{1}{2}x_n^2$. Prove that for $n \geqslant 3$, we have

$$|x_n - \sqrt{2}| < \left(\frac{1}{2}\right)^n.$$

8. Assume $a > 0$, show that

$$\sqrt{a + \sqrt{2a + \sqrt{3a + \cdots + \sqrt{na}}}} < \sqrt{a} + 1.$$

9. Let $a_1, a_2, \cdots, a_n$ be positive real numbers with product 1. Prove that

$$\sum_{i=1}^{n} a_i \geqslant n.$$

10. Assume a sequence $\{a_n\}$ satisfies: $a_1 = a_2 = 1$, $a_{n+2} = a_{n+1} + a_n$. Denote S_n the sum of the first n terms in the sequence $\{a_n\}$, show that

$$S_n = \sum_{k=1}^{n} \frac{a_k}{2^k} < 2.$$

11. Assume $r_1, r_2, \cdots, r_n$ are real numbers $\geqslant 1$. Prove that

$$\frac{1}{r_1+1} + \frac{1}{r_2+1} + \cdots + \frac{1}{r_n+1} \geqslant \frac{n}{\sqrt[n]{r_1 r_2 \cdots r_n} + 1}.$$

12. Assume $f(n)$ is defined on positive integers and it satisfies: $f(1) = 2$, $f(n+1) = f^2(n) - f(n) + 1$, $n = 1, 2, \cdots$. Prove that for any integer $n > 1$, we have

$$1 - \frac{1}{2^{2^{n-1}}} < \frac{1}{f(1)} + \frac{1}{f(2)} + \cdots + \frac{1}{f(n)} < 1 - \frac{1}{2^{2^n}}.$$

13. Assume $\mathbf{N}_+$ is the set of positive integers, $\mathbf{R}$ is the set of real numbers. S is a set of functions f: $\mathbf{N}_+ \to \mathbf{R}$ that also satisfy the following two conditions:

(1) $f(1) = 2$.

(2) $f(n+1) \geqslant f(n) \geqslant \frac{n}{n+1} f(2n)$ $(n = 1, 2, \cdots)$.

Find the minimum positive integer M, such that for any $f \in S$ and any $n \in \mathbf{N}_+$, we have $f(n) < M$.

14. Let $a_1, a_2, \cdots, a_n$ be non-negative real numbers satisfying $\sum_{i=1}^{n} a_i = 4$, $n \geqslant 3$. Prove that

$$a_1^3 a_2 + a_2^3 a_3 + \cdots + a_{n-1}^3 a_n + a_n^3 a_1 \leqslant 27.$$

15. Assume n is a positive integer and x is a positive real number. Show that

$$\sum_{k=1}^{n} \frac{x^{k^2}}{k} \geqslant x^{\frac{1}{2}n(n+1)}.$$

16. Assume $z_i\,(1 \leqslant i \leqslant n)$ are n complex numbers, $s_i = z_1 + z_2 + \cdots + z_i$, $1 \leqslant i \leqslant n$. Prove that

$$\sum_{1 \leqslant i < j \leqslant n} |s_j - z_i| \leqslant \sum_{k=1}^{n} [(n+1-k)|z_k| + (k-2)|s_k|].$$

Chapter 8 Inequality and Extremum for Multi-Variable Function

In this chapter we study the method of fixed variables. When we try to solve an extremum problem that is complicated by many variables, we can fix some of the variables first and alter the rest of the variables so we can understand their relationship with the original quantity. Then, we "unfreeze" those previously fixed variables and study their effects until we solve the problem.

Of course, when we adopt the method of fixed variables, we can start with the function value and let the function value approach its extremum step by step; or we can focus on the variables directly and find out when do they reach extremum. We can handle the problem directly and we can also introduce the method of contradiction. Based on different approaches, we discuss in turn the iterated method of finding extremum, the mollification method and the adjustment method.

8.1 Iterated Method of Finding Extremum

The idea of using the iterated method to find the extremum is to find the extremum of an extremum: we fix some variables first and find the extremum with the rest of the variables, then we relax those fixed variables and estimated the extremum we have just acquired. In turn, we will finally reach the extremum for the original quantity. The Iterated Method of Finding Extremum is also referred to as the "Successive Approaching Method" and requires skillful handling.

Eg. 1 Find the minimum of a three-digit number divided by the sum of its digits.

Soln. Assume the three-digit number is $100x + 10y + z$, where x,

y, z are positive integers and $1 \leqslant x \leqslant 9$, $0 \leqslant y, z, \leqslant 9$.

Hence, the ratio we are interested in is

$$f(x, y, z) = \frac{100x + 10y + z}{x + y + z} = 1 + \frac{99x + 9y}{x + y + z}. \quad ①$$

On the left hand side of ①, only the denominator contains variable z. So when x, y are fixed, $f(x, y, z)$ reaches its minimum if and only if $z = 9$.

Similarly, we have

$$\begin{aligned} f(x, y, z) &\geqslant 1 + \frac{99x + 9y}{9 + x + y} = 10 + \frac{90x - 81}{9 + x + y} \\ &\geqslant 10 + \frac{90x - 81}{18 + x} = 100 - \frac{1701}{18 + x} \\ &\geqslant 100 - \frac{1701}{18 + 1} = \frac{199}{19}. \end{aligned}$$

Thus, $f_{\min}(x, y, z) = f(1, 9, 9) = \dfrac{199}{19}$.

Eg. 2 Let A, B, C be non-negative real numbers such that $A + B + C = \frac{\pi}{2}$. Write $M = \sin A + \sin B + \sin C$, $N = \sin^2 A + \sin^2 B + \sin^2 C$. Prove that $M^2 + N \leqslant 3$.

Proof. WLOG, assume $C \leqslant A, B$, then $C \in [0, \pi/6]$,

$$\begin{aligned} M &= 2\sin\frac{A + B}{2}\cos\frac{A - B}{2} + \sin C \\ &= 2\sin\left(\frac{\pi}{4} - \frac{C}{2}\right)\cos\frac{A - B}{2} + \sin C, \end{aligned}$$

$$N = \frac{1}{2}(1 - \cos 2A) + \frac{1}{2}(1 - \cos 2B) + \sin^2 C.$$

Fixing C, we have

$$\begin{aligned} M^2 + N = {} & 1 + 2\sin^2 C + 4\sin C \cdot \sin\left(\frac{\pi}{4} - \frac{C}{2}\right)\cos\frac{A - B}{2} \\ & + 4\sin^2\left(\frac{\pi}{4} - \frac{C}{2}\right) \cdot \cos^2\frac{A - B}{2} - \sin C \cdot \cos(A - B) \end{aligned}$$

$$= 1 + 2\sin^2 C + 4\sin C \cdot \sin\left(\frac{\pi}{4} - \frac{C}{2}\right)\cos\frac{A-B}{2}$$

$$+ 4\sin^2\left(\frac{\pi}{4} - \frac{C}{2}\right) \cdot \cos^2\frac{A-B}{2} + \sin C$$

$$- 2\sin C\cos^2\frac{A-B}{2}$$

$$= 1 + 2\sin^2 C + \sin C + 4\sin\left(\frac{\pi}{2} - \frac{C}{2}\right)\sin C$$

$$+ \left[4\sin^2\left(\frac{\pi}{4} - \frac{C}{2}\right) - 2\sin C\right]\cos^2\frac{A-B}{2}$$

$$= 1 + 2\sin^2 C + \sin C + 4\sin\left(\frac{\pi}{4} - \frac{C}{2}\right)\sin C$$

$$+ 2(1 - 2\sin C)\cos^2\frac{A-B}{2}$$

$$\leqslant 1 + 2\sin^2 C + \sin C + 4\sin\left(\frac{\pi}{4} - \frac{C}{2}\right)\sin C$$

$$+ 2(1 - 2\sin C)$$

$$= 3 + 2\sin^2 C - 3\sin C + 4\sin\left(\frac{\pi}{4} - \frac{C}{2}\right)\sin C.$$

Denote $t = \frac{\pi}{4} - \frac{C}{2}$, then $t \in \left[\frac{\pi}{6}, \frac{\pi}{4}\right]$, thus:

$$\begin{aligned} M^2 + N &= 3 + 2\cos^2 2t - 3\cos 2t + 4\sin t\cos 2t \\ &= 3 + \cos 2t(2\cos 2t - 3 + 4\sin t) \\ &= 3 + \cos 2t(-4\sin^2 t + 4\sin t - 1) \\ &= 3 - \cos 2t(2\sin t - 1)^2 \\ &\leqslant 3. \end{aligned}$$

When $A = B = \frac{\pi}{6}$ or $\frac{\pi}{4}$, (i.e. when $A = B = C = \frac{\pi}{6}$ or $A = B = \frac{\pi}{4}$, $C = 0$) the equality holds.

Eg. 3 Assume A, B, C, D are 4 points in space. At most one length of segments AB, AC, AD, BC, BD, CD is greater than 1. Find the maximum for the sum of these 6 lengths.

Soln. Assume AD is the longest among the 6.

(1) Fix all other 5 segments. It's easy to see that when A and D are the two opposite vertices of the parallelogram $ABCD$, AD can reach its maximum length.

(2) Fix points B and C. Then A and D must lie in the intersection of the two unit circles with centers B and C respectively. In this case, the maximum distance we can have is when A and D are exactly the two intersection points of the two circles, then AB, BD, AC, CD all reach a maximum of 1.

(3) We see that when BC decreases, AD increases and when BC increases, AD decreases. So instead we try to find out the maximum for $BC + AD$ when the other 4 sides have fixed length 1.

Denote $\angle ABO = \theta$. Since $AB = 1$, $0 < BC \leqslant 1$, we know $\theta \in \left[\frac{\pi}{3}, \frac{\pi}{2}\right)$.

Hence, $AD + BC = 2(\sin\theta + \cos\theta) = 2\sqrt{2}\sin\left(\theta + \frac{\pi}{4}\right)$ and is maximized when $\theta = \frac{\pi}{3}$.

At this time, the sum of the 6 lengths is

$$4 + 2\left(\sin\frac{\pi}{3} + \cos\frac{\pi}{3}\right) = 5 + \sqrt{3}.$$

Eg. 4 Let a, b, c be non-negative real numbers such that $a + b + c = 1$. Find the maximum for $S = ab + bc + ca - 3abc$.

Soln. WLOG, assume $a \geqslant b \geqslant c$, then $c \leqslant \frac{1}{3}$.

Fix c,

$$\begin{aligned} S &= ab + bc + ca - 3abc \\ &= ab(1 - 3c) + c(a + b). \end{aligned}$$

Since $a + b$ is fixed, $1 - 3c \geqslant 0$, ab is maximized when $a = b$, so S is also maximized when $a = b$.

Adjust a, b, c to $a = b \geqslant c$, then $b \geqslant \frac{1}{3}$.

Now fix b.

$$S = ab + bc + ca - 3abc$$
$$= ac(1-3b) + b(a+c).$$

Since $1-3b \leqslant 0$, $a+c$ is fixed, so S is maximized when $a-c$ is maximized. However, $a-c \leqslant 1-b$ and the equality holds when $c=0$.

As a result, we only need to find the maximum for ab given $c=0$ and $a+b=1$, which is clearly $\frac{1}{4}$.

Therefore, when one of a, b, c is 0 and the other two $\frac{1}{2}$, S reaches its maximum $\frac{1}{4}$.

Eg. 5 Let α, β, γ be non-negative numbers satisfy $\alpha+\beta+\gamma=\frac{\pi}{2}$. Find the minimum for

$$f(\alpha, \beta, \gamma) = \frac{\cos\alpha\cos\beta}{\cos\gamma} + \frac{\cos\beta\cos\gamma}{\cos\alpha} + \frac{\cos\gamma\cos\alpha}{\cos\beta}.$$

Soln. WLOG, assume $\gamma \leqslant \alpha, \beta$, then $\gamma \in \left[0, \frac{\pi}{6}\right]$.

Fix γ. Since

$$\frac{\cos\beta\cos\gamma}{\cos\alpha} + \frac{\cos\gamma\cos\alpha}{\cos\beta} = 2\cos\gamma\left(\frac{\cos^2\gamma}{\sin\gamma + \cos(\alpha-\beta)} + \sin\gamma\right),$$

$$\cos\alpha \cdot \cos\beta = \frac{1}{2}(\sin\gamma + \cos(\alpha-\beta)).$$

We have

$$f = \frac{1}{2}\,\frac{\sin\gamma + \cos(\alpha-\beta)}{\cos\gamma} + 2\cos\gamma\left(\sin\gamma + \frac{\cos^2\gamma}{\sin\gamma + \cos(\alpha-\beta)}\right)$$
$$= \sin 2\gamma + \frac{1}{2}\,\frac{\sin\gamma + \cos(\alpha-\beta)}{\cos\gamma} + 2 \cdot \frac{\cos^3\gamma}{\sin\gamma + \cos(\alpha-\beta)}.$$

Since $\sin\gamma + \cos(\alpha-\beta) \leqslant \sin\gamma + 1 \leqslant \frac{3}{2} \leqslant 2\cos^2\gamma$, we have

$$f \geqslant \sin 2\gamma + \frac{1}{2}\frac{1+\sin\gamma}{\cos\gamma} + 2\cdot\frac{\cos^3\gamma}{1+\sin\gamma}$$
$$= 2\cos\gamma + \frac{1}{2}\frac{1+\sin\gamma}{\cos\gamma}$$
$$= 2\cos\gamma + \frac{1}{2}\cot\left(\frac{\pi}{4}-\frac{\gamma}{2}\right).$$

Let $\theta = \frac{\pi}{4} - \frac{\gamma}{2} \in \left[\frac{\pi}{6}, \frac{\pi}{4}\right]$, thus,

$$f = 2\sin 2\theta + \frac{1}{2}\cot\theta = \frac{4\tan^2\theta}{1+\tan^2\theta} + \frac{1}{2}\cot\theta$$
$$= \frac{5}{2} + \frac{1}{2}\frac{(1-\tan\theta)(5\tan^2\theta - 4\tan\theta + 1)}{\tan\theta(1+\tan^2\theta)}$$
$$\geqslant \frac{5}{2}\ \left(\tan\theta \in \left[\frac{\sqrt{3}}{3}, 1\right]\right).$$

Therefore, when $\alpha = \beta = \frac{\pi}{4}$, $\gamma = 0$, f reaches its minimum $\frac{5}{2}$.

8.2 The Mollification Method

In the previous section, we use the iterated method to find the extremum. During each iteration, we have to guarantee that the equality holds for the same set of variables: this often requires us to find the conditions for reaching the extremum first. Now we can use another method in dealing with the extremum problems. When we know (or have guessed) when the equality holds, we don't need to make sure that in each step the equality holds for the same set of variables anymore. Instead, we "push" the variable array $(x_1, x_2, \cdots, x_n)$ closer and closer to the point where extremum is reached and we only need to guarantee that we execute finite many of these steps. We call this method the "Mollification Method".

When we use the mollification method, we have to be careful so that the function value move towards the extremum under mollification.

For example, if we want to find the maximum, then the mollification should not decrease the function value, sometimes this requires us to pick a suitable set of variables.

Eg. 6 Let x, y, z be all non-negative real numbers such that $x + y + z = 1$. Prove that

$$yz + zx + xy - 2xyz \leqslant \frac{7}{27}.$$

Proof. It's easy to see that when $x = y = z = \frac{1}{3}$, the equality holds. WLOG assume $x \geqslant y \geqslant z$, then $x \geqslant \frac{1}{3} \geqslant z$.

Let $x' = \frac{1}{3}$, $y' = y$, $z' = x + z - \frac{1}{3}$, thus:

$$x' + z' = x + z,\ x' \cdot z' \geqslant x \cdot z.$$

Hence,

$$\begin{aligned} yz + zx + xy - 2xyz &= y(x + z) + (1 - 2y)xz \\ &\leqslant y'(x' + z') + (1 - 2y')x'z' \\ &= \frac{1}{3}(y' + z') + \frac{1}{3}y'z' \leqslant \frac{2}{9} + \frac{1}{27} = \frac{7}{27}. \end{aligned}$$

Eg. 7 Let $x_1, x_2, \cdots, x_n (n \geqslant 3)$ be non-negative real numbers satisfy the inequality

$$x_1 + x_2 + \cdots + x_n \leqslant \frac{1}{2}.$$

Find the minimum for $(1 - x_1)(1 - x_2)\cdots(1 - x_n)$.

Soln. When $x_1, x_2, \cdots, x_{n-2}, x_{n-1} + x_n$ are all fixed, since

$$(1 - x_{n-1})(1 - x_n) = 1 - (x_{n-1} + x_n) + x_{n-1}x_n.$$

We see that the larger $| x_{n-1} - x_n |$ gets, the smaller is the above expression.

Thus, when $n \geqslant 3$, denote

$$x_i' = x_i,\ i = 1, 2, \cdots, n - 2,\ x_{n-1}' = x_{n-1} + x_n,\ x_n' = 0. \quad ①$$

Hence,

$$x'_{n-1}+x'_n=x_{n-1}+x_n,\ x'_{n-1}x'_n=0\leqslant x_{n-1}x_n.$$

We have:

$$(1-x_1)(1-x_2)\cdots(1-x_n)\geqslant(1-x'_1)(1-x'_2)\cdots(1-x'_{n-1}),$$

where $x'_1+x'_2+\cdots+x'_{n-1}=x_1+x_2+\cdots+x_n\leqslant\frac{1}{2}$.

Perform mollification as ① for another $n-2$ times, we obtain

$$(1-x_1)(1-x_2)\cdots(1-x_n)\geqslant 1-(x_1+x_2+\cdots+x_n)\geqslant\frac{1}{2},$$

the equality holds when $x_1=\frac{1}{2},\ x_2=x_3=\cdots=x_n=0$.

Eg. 8 Let $a,\ b,\ c,\ d$ be non-negative real numbers and $a+b+c+d=1$. Prove that

$$bcd+cda+dab+abc\leqslant\frac{1}{27}+\frac{176}{27}abcd.$$

Proof. If $d=0$, then $abc\leqslant\frac{1}{27}$, the inequality obviously holds. If $a,\ b,\ c,\ d>0$, we only need to show that

$$f(a,\ b,\ c,\ d)=\sum\frac{1}{a}-\frac{1}{27}\frac{1}{abcd}\leqslant\frac{176}{27}.$$

Let

$$a\leqslant\frac{1}{4}\leqslant b,\ a'=\frac{1}{4},\ b'=a+b-\frac{1}{4},$$

then

$$\begin{aligned}
&f(a,\ b,\ c,\ d)-f(a',\ b',\ c',\ d')\\
=&\left(\frac{1}{a}+\frac{1}{b}-\frac{1}{a'}-\frac{1}{b'}\right)-\frac{1}{27}\cdot\frac{1}{cd}\cdot\left(\frac{1}{ab}-\frac{1}{a'b'}\right)\\
=&\frac{a'b'-ab}{aba'b'}(a+b)\left(1-\frac{1}{27}\frac{1}{cd(a+b)}\right)\\
\leqslant&\frac{a'b'-ab}{aba'b'}(a+b)\left[1-\frac{1}{27}\cdot\left(\frac{a+b+c+d}{3}\right)^{-3}\right]\\
=&0.
\end{aligned}$$

Thus $f(a, b, c, d) \leqslant f(a', b', c, d)$.

Via at most two mollifications, we have

$$\begin{aligned} f(a, b, c, d) &\leqslant f\left(\frac{1}{4}, a+b-\frac{1}{4}, c, d\right) \\ &\leqslant f\left(\frac{1}{4}, \frac{1}{4}, a+b+c-\frac{1}{2}, d\right) \\ &\leqslant f\left(\frac{1}{4}, \frac{1}{4}, \frac{1}{4}, \frac{1}{4}\right) \\ &= \frac{176}{27}. \end{aligned}$$

Hence the original inequality holds.

Eg. 9 Show that for all n -polygons with fixed perimeter l, the regular one has the maximum surface area.

Proof. (1) First, concave polygons cannot have the maximum area.

In Figure 8.1, assume $A_1A_2\cdots A_n$ is a concave polygon. If we "fold out" $\Delta A_{i-1}A_iA_{i+1}$ to $\Delta A_{i-1}A'_iA_{i+1}$ (i.e. use the middle point of $A_{i+1}A_{i-1}$ as the center, $\Delta A_{i-1}A'_iA_{i+1}$ is center symmetric to $\Delta A_{i-1}A_iA_{i+1}$), we obtain a convex polygon $A_1\cdots A_{i-1}A'_iA_{i+1}\cdots A_n$ with perimeter l and larger area. Hence, in the following discussion we only consider convex polygons.

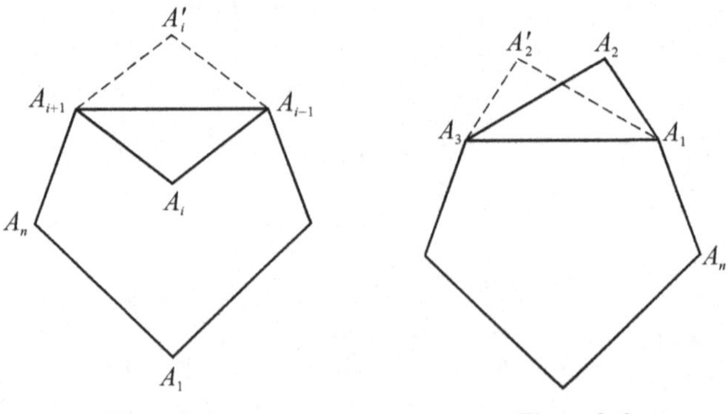

Figure 8.1 Figure 8.2

(2) Assume the convex polygon $A_1A_2\cdots A_n$ is not equilateral and assume there exist two adjacent sides, one less than $\frac{l}{n}$ and the other

greater than $\frac{l}{n}$. WLOG, assume these two sides are A_1A_2 and A_2A_3 respectively. Now connect A_1A_3 and construct a triangle $\Delta A_1A'_2A_3$ using A_1A_3, $\frac{l}{n}$ and $A_1A_2+A_2A_3-\frac{l}{n}$ as the length of its three sides. (as shown in Figure 8.2)

Thus, $A_1A_2 < A_1A'_2$, $A'_2A_3 < A_2A_3$, and $A_1A'_2+A'_2A_3 = A_1A_2 + A_2A_3$.

From Heron's Formula, we know that $S_{\triangle A_1A'_2A_3} > S_{\triangle A_1A_2A_3}$, hence the area of $A_1A'_2A_3\cdots A_n$ is greater than the area of $A_1A_2\cdots A_n$, and the former has a side $A_1A'_2$ of length $\frac{l}{n}$.

If the polygon $A_1A'_2A_3\cdots A_n$ is still not equilateral, we perform similar rounds of mollification such that each time we obtain at least one more side with length $\frac{l}{n}$. Hence, we will reach to an equilateral polygon after at most $n-1$ mollifications. In other words, non-equilateral n-polygon must have smaller area than some equilateral polygon with the same perimeter.

(3) Assume the convex polygon $A_1A_2\cdots A_n$ is not equilateral and there exist two sides $A_iA_{i+1} < \frac{l}{n} < A_jA_{j+1}$ that are not adjacent. Connect A_iA_j and draw MN, the perpendicular bisector of A_iA_j. Now set MN as the axis of symmetry and reflect the polygon $A_iA_{i+1}\cdots A_j$ to the polygon $A_iA'_{i+1}\cdots A_j$, as shown in Figure 8.3.

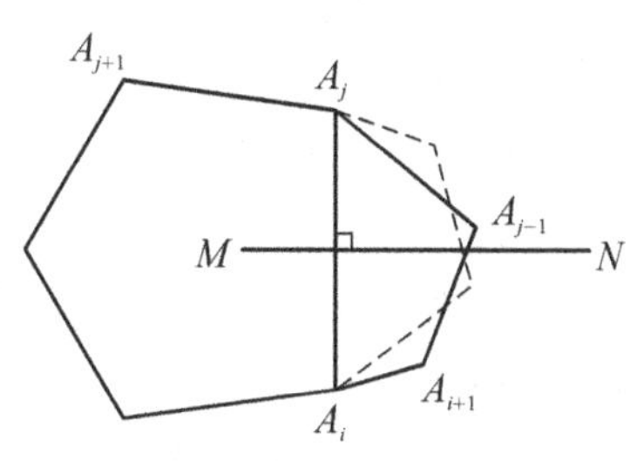

Figure 8.3

Thus, the new n-polygon has same perimeter and area as the original n-polygon.

However, the new polygon has two adjacent sides; one side is greater than $\frac{l}{n}$ while the other is smaller than $\frac{l}{n}$. This is exactly case (2).

We conclude that any non-equilateral n-polygon with perimeter l

must have smaller area than some equilateral n -polygon with perimeter l .

(4) It is well known that for an equilateral n -polygon, the regular one has the largest area.

To sum, for all n -polygon with fixed perimeter l , the regular one has the largest area.

8.3 The Adjustment Method

For some extremum problems such as the discrete extremum problem, there are only finite many situations and naturally there exist maximum and minimum. As a result, we can make use of the existence of the extremum and adopt the adjustment method along with the method of contradiction to solve the problem. Since the extremum must exist, we don't need to make sure that each step reaches extremum as the iterated method we discussed before, nor do we need to approach the extremum as the mollification method. Rather, we make adjustments in separate cases and only need to show that these cases cannot reach extremum.

Eg. 10 Several positive integers add up to 2005. Find the maximum for their product.

Soln. Since there are finitely many groups of positive integers that add up to 2005, there must be a maximum for their products.

Assume x_1, x_2, $\cdots$, x_n are positive integers, $x_1+x_2+\cdots+x_n=2005$ and the product $u=x_1x_2\cdots x_n$ reaches its maximum. Thus:

(1) $x_i\leqslant 4(i=1, 2, \cdots, n)$. Otherwise, if some $x_j>4$, since $2+(x_j-2)=x_j$, $2(x_j-2)=x_j+(x_j-4)>x_j$, we can replace x_j with 2 and x_j-2 so that the sum doesn't change but the product increases. This is a contradiction!

(2) $x_j\geqslant 2(i=1, 2, \cdots, n)$. Otherwise say some $x_k=1$, since $x_k\cdot x_i=x_i<x_i+x_k$, so we can replace x_kx_i with x_k+x_i to increase the product while keeping the sum. Also a contradiction!

(3) Since $4=2+2=2\times 2$, hence $x_j=4$ can be replaced by two 2's while keeping u unchanged.

(4) Assume now $u = 2^{\alpha} \cdot 3^{\beta}$. If $\alpha > 3$, then $u = 2^3 \cdot 2^{\alpha-3} \cdot 3^{\beta}$. Note that $2+2+2=3+3$, so the sum is not changed while u increases, this leads to a contradiction!

To conclude, since $2005 = 3+1$, we have $u_{\max} = 2^2 \times 3^{667}$.

In general, if the sum is m, then

$$u_{\max} = \begin{cases} 3^s, & \text{if } m = 3s, \\ 2^2 \cdot 3^{s-1}, & \text{if } m = 3s+1, \\ 2 \cdot 3^s, & \text{if } m = 3s+2. \end{cases}$$

Eg. 11 Assume $a_1, a_2, \cdots, a_n, \cdots$ is a non-decreasing sequence of positive integers. For $m \geqslant 1$, define $b_m = \min\{n, a_n \geqslant m\}$, i.e. b_m is the smallest n such that $a_n \geqslant m$. Given that $a_{19} = 85$, find the maximum for

$$a_1 + a_2 + \cdots + a_{19} + b_1 + b_2 + \cdots + b_{85}. \quad ①$$

Soln. If there exists i, such that $a_i < a_{i+1} (1 \leqslant i \leqslant 18)$, we perform the following adjustment: $a'_i = a_i + 1$, $a'_j = a_j (j \neq i)$. We denote the adjusted b_j as b'_j $(j = 1, 2, \cdots, 85)$.
From the definition we know that

$$b_{a_i+1} = i+1,$$
$$b'_{a_i+1} = i = b_{a_i+1} - 1,$$
$$b'_j = b_j (j \neq a_i + 1).$$

Thus, the above adjustment decreases b_{a_i+1} by 1 and maintains the rest of the b_j's, hence such adjustment does not change the value of ①.

Now, we can perform a series of adjustments such that $a_1 = a_2 = \cdots = a_{19} = 85$ and keeps the value of ①. However at this time $b_1 = b_2 = \cdots = b_{85} = 1$, so the maximum for ① is

$$19 \times 85 + 1 \times 85 = 20 \times 85 = 1700.$$

Exercise 8

1. Assume that $x \geqslant 0$, $y \geqslant 0$, $z \geqslant 0$, $x + y + z = 1$. Find the maximum and minimum for $S = 2x^2 + y + 3z^2$.

2. Prove that in $\triangle ABC$, for $0 \leqslant \lambda_i \leqslant 2 (i = 1, 2, 3)$, we have

$$\sin \lambda_1 A + \sin \lambda_2 B + \sin \lambda_3 C \leqslant 3 \sin \frac{\lambda_1 A + \lambda_2 B + \lambda_3 C}{3}.$$

3. Let $\theta_1, \theta_2, \cdots, \theta_n$ be non-negative and $\theta_1 + \theta_2 + \cdots + \theta_n = \pi$. Find the maximum for

$$\sin^2 \theta_1 + \sin^2 \theta_2 + \cdots + \sin^2 \theta_n.$$

4. Assume $a, b, c, d \geqslant 0$, and $a + b + c + d = 4$. Prove that

$$bcd + cda + dab + abc - abcd \leqslant \frac{1}{2}(ab + ac + ad + bc + bd + cd).$$

5. Let x_i be non-negative real numbers, $i = 1, 2, 3, 4$, and

$$x_1 + x_2 + x_3 + x_4 = 1.$$

Denote

$$S = 1 - \sum_{i=1}^{4} x_i^3 - 6 \sum_{1 \leqslant i < j < k \leqslant 4} x_i x_j x_k.$$

Find the range for S.

6. Let $x_1, x_2, \cdots, x_n$ be n non-negative real numbers ($n > 2$, $n \in \mathbf{N}_+$), and

$$\sum_{i=1}^{n} x_i = n, \ \sum_{i=1}^{n} i x_i = 2n - 2.$$

Find the maximum for $x_1 + 4x_2 + \cdots + n^2 x_n$.

7. For non-negative real numbers $x_i (i = 1, 2, \cdots, n)$ that satisfy $x_1 + x_2 + \cdots + x_n = 1$, find the maximum for $\sum_{j=1}^{n} (x_j^4 - x_j^5)$.

8. Let a_1, a_2, a_3 be non-negative real numbers. Prove that

$$a_1 + a_2 + a_3 + 3\sqrt[3]{a_1 a_2 a_3} \geqslant 2(\sqrt{a_1 a_2} + \sqrt{a_2 a_3} + \sqrt{a_3 a_1}).$$

9. Assume $a_1, a_2, \cdots, a_{10}$ are 10 mutually different positive integers that sum up to 2002. Find the minimum for $a_1a_2 + a_2a_3 + \cdots + a_{10}a_1$.

10. Let a, b, c be non-negative real numbers satisfy $ab + bc + ca = 1$, find the minimum for $\frac{1}{a+b} + \frac{1}{b+c} + \frac{1}{c+a}$.

11. Let $a_1, a_2, \cdots, a_{2001}$ be non-negative real numbers such that:

(1) $a_1 + a_2 + \cdots + a_{2001} = 2$.

(2) $a_1a_2 + a_2a_3 + \cdots + a_{2000}a_{2001} + a_{2001}a_1 = 1$.

Find the extremum for $S = a_1^2 + a_2^2 + \cdots + a_{2001}^2$.

12. Let $x_1, x_2, \cdots, x_n$ be non-negative real numbers with sum 1. Find the maximum for the sum

$$\sum_{1 \leqslant i < j \leqslant n} x_i x_j (x_i + x_j).$$

13. Assume n a fixed integer, $n \geqslant 2$.

(1) Determine the minimum constant c, such that the inequality

$$\sum_{1 \leqslant i < j \leqslant n} x_i x_j (x_i^2 + x_j^2) \leqslant c \left(\sum_{1 \leqslant i \leqslant n} x_i \right)^4$$

holds for all non-negative real numbers $x_1, x_2, \cdots, x_n$.

(2) For the constant you find in (1), determine when does the equality hold.

14. Denote $S = \left\{ \frac{l}{1997} \,\middle|\, l = 0, 1, 2, \cdots, 1996 \right\}$. Three numbers x, y, z in S satisfy $x^2 + y^2 - z^2 = 1$. Find the minimum and maximum for $x + y + z$.

Chapter 9 Special Techniques for Proving Inequalities

Inequality is a field of great depth. It often encourages us to try different methods and rewards us with creative thinking. During the previous chapters, we have already studied many aspects and went through an impressive variety of inequalities. In this final chapter, we present a few more special techniques that may prove helpful when we have exhausted our bag of tricks, or have found traditional approaches somewhat ineffective.

9.1 Truncated Sums

Sometimes, if estimating the sum on one side of an inequality proves difficult, we can break the summands into two parts and deal with each part separately. This clearly involves a lot of deliberation, as we will be facing the task of finding a suitable break point as well as customized treatment for each individual component of the sum.

Eg. 1 Prove that for any $n \in \mathbf{N}_+$,

$$\left|\sum_{k=1}^{n}(-1)^k\left\{\frac{n}{k}\right\}\right| \leqslant 3\sqrt{n}$$

where $\{t\} = t - \lfloor t \rfloor$ denotes the fractional part of t.

Hint. For a sum of monotonically decreasing series with alternating signs, the following inequality is easy to prove: Assume $a_1 > a_2 > \cdots > a_k > 0$, then

$$-a_1 < a_1 - a_2 + a_3 - a_4 + \cdots < a_1. \quad ①$$

There are two possible ways to estimate the left hand side:

(1) Trivially, the absolute value of each term in the sum is $\leqslant 1$, hence LHS $\leqslant n$.

(2) Applying ①, we have

$$\left|\sum_{k=1}^{n}(-1)^{k}\left\{\frac{n}{k}\right\}\right| = \left|\sum_{k=1}^{n}(-1)^{k}\frac{n}{k} - \sum_{k=1}^{n}(-1)^{k}\left\lfloor\frac{n}{k}\right\rfloor\right|$$
$$\leqslant \left|\sum_{k=1}^{n}(-1)^{k}\frac{n}{k}\right| + \left|\sum_{k=1}^{n}(-1)^{k}\left\lfloor\frac{n}{k}\right\rfloor\right|$$
$$\leqslant n + n = 2n. \qquad ②$$

Surprisingly, (2) gives us an even looser estimate! The reason of this can be attributed to the fact that the corresponding term a_1 in ① is too large. We will be better off to start with a smaller a_1. On the other hand, we do not want to use the trivial estimate in (1) on too many terms. These two findings suggest to break the sum into two parts.

Proof.

$$\left|\sum_{k=A}^{n}(-1)^{k}\frac{n}{k} - \sum_{k=A}^{n}(-1)^{k}\left\lfloor\frac{n}{k}\right\rfloor\right| \leqslant \left|\sum_{k=A}^{n}(-1)^{k}\frac{n}{k}\right| + \left|\sum_{k=A}^{n}(-1)^{k}\left\lfloor\frac{n}{k}\right\rfloor\right|$$
$$\leqslant \frac{n}{A} + \left\lfloor\frac{n}{A}\right\rfloor \leqslant \frac{2n}{A},$$

where A is the break-point. As a result,

$$\text{LHS} \leqslant \left|\sum_{k\leqslant A-1}^{n}(-1)^{k}\left\{\frac{n}{k}\right\}\right| + \left|\sum_{k\geqslant A}^{n}(-1)^{k}\left\{\frac{n}{k}\right\}\right|$$
$$\leqslant \left|\sum_{k\leqslant A-1}^{n}(-1)^{k}\left\{\frac{n}{k}\right\}\right| + \left|\sum_{k\geqslant A}^{n}(-1)^{k}\frac{n}{k}\right| + \left|\sum_{k\geqslant A}^{n}(-1)^{k}\left\lfloor\frac{n}{k}\right\rfloor\right|$$
$$\leqslant (A-1) + \frac{2n}{A}.$$

The last step of the above argument have used method (1) and (2) alternatively.

Use $A = \lfloor\sqrt{2n}\rfloor + 1$, LHS $\leqslant \sqrt{2n} + \sqrt{2n} < 3\sqrt{n}$.

Eg. 2 Real numbers $a_1, a_2, \cdots, a_n (n \geqslant 3)$ satisfy $a_1 + a_2 + \cdots + a_n = 0$, and

$$2a_k \leqslant a_{k-1} + a_{k+1},\ k = 2, 3, \cdots, n-1.$$

Find the smallest $\lambda(n)$, such that for all $k \in \{1, 2, \cdots, n\}$, we have

$$|a_k| \leqslant \lambda(n) \cdot \max\{|a_1|, |a_n|\}.$$

(Chinese Western Mathematical Olympiad, 2009)

Soln. First, set $a_1 = 1$, $a_2 = -\frac{n+1}{n-1}$, $a_k = -\frac{n+1}{n-1} + \frac{2n(k-2)}{(n-1)(n-2)}$, $k = 2, \cdots, n-1$.

The first constraint $a_1 + \cdots + a_n = 0$ holds and so does the second constraint $2a_k \leqslant a_{k-1} + a_{k+1}$, $k = 2, 3, \cdots, n-1$. Thus

$$\lambda(n) \geqslant \frac{n+1}{n-1}.$$

We prove in the following that when $\lambda(n) = \frac{n+1}{n-1}$, for all $k \in \{1, 2, \cdots, n\}$,

$$|a_k| \leqslant \lambda(n) \cdot \max\{|a_1|, |a_n|\}.$$

Since $2a_k \leqslant a_{k-1} + a_{k+1}$, $a_{k+1} - a_k \geqslant a_k - a_{k-1}$. Therefore,

$$a_n - a_{n-1} \geqslant a_{n-1} - a_{n-2} \geqslant \cdots \geqslant a_2 - a_1.$$

Hence,

$$\begin{aligned}(k-1)(a_n - a_1) &= (k-1)[(a_n - a_{n-1}) + (a_{n-1} - a_{n-2}) + \cdots + (a_2 - a_1)] \\ &\geqslant (n-1)[(a_k - a_{k-1}) + (a_{k-1} - a_{k-2}) + \cdots + (a_2 - a_1)] \\ &= (n-1)(a_k - a_1).\end{aligned}$$

It follows that

$$a_k \leqslant \frac{k-1}{n-1}(a_n - a_1) + a_1 = \frac{1}{n-1}[(k-1)a_n + (n-k)a_1]. \quad ①$$

Similar to ①, for fixed k, $k \neq 1, n$, when $1 \leqslant j \leqslant k$,

$$a_j \leqslant \frac{1}{k-1}[(j-1)a_k + (k-j)a_1].$$

When $k \leqslant j \leqslant n$,

$$a_j \leqslant \frac{1}{n-k}[(j-k)a_n + (n-j)a_k].$$

As a result,

$$\sum_{j=1}^{k} a_j \leqslant \frac{1}{k-1}\sum_{j=1}^{k}[(j-1)a_k + (k-j)a_1] = \frac{k}{2}(a_1 + a_k),$$

$$\sum_{j=k}^{n} a_j \leqslant \frac{1}{n-k}\sum_{j=k}^{n}[(j-k)a_n + (n-j)a_k] = \frac{n+1-k}{2}(a_k + a_n).$$

We add the above two formula to obtain

$$\begin{aligned} a_k &= \sum_{j=1}^{k} a_j + \sum_{j=k}^{n} a_j \leqslant \frac{k}{2}(a_1 + a_k) + \frac{n+1-k}{2}(a_k + a_n) \\ &= \frac{k}{2}a_1 + \frac{n+1}{2}a_k + \frac{n+1-k}{2}a_n. \end{aligned}$$

This yields

$$a_k \geqslant -\frac{1}{n-1}[ka_1 + (n+1-k)a_n]. \qquad ②$$

Combining ① and ②, we have

$$\begin{aligned} |a_k| &\leqslant \max\left\{\frac{1}{n-1}|(k-1)a_n + (n-k)a_1|, \frac{1}{n-1}|ka_1 + (n+1-k)a_n|\right\} \\ &\leqslant \frac{n+1}{n-1}\max\{|a_1|, |a_n|\}, \ k = 2, 3, \cdots, n-1. \end{aligned}$$

From the discussion above, the smallest $\lambda(n)$ is $\frac{n+1}{n-1}$.

Eg. 3 Let $a_1, a_2, \cdots, a_n$ be positive real numbers such that $\sum_{i=1}^{n} a_i = 1$.

Denote k_i the number of a_j's satisfying $\frac{1}{2^i} \leqslant a_j < \frac{1}{2^{i-1}}$. Prove that

$$\sum_{i=1}^{+\infty}\sqrt{\frac{k_i}{2^i}} \leqslant \sqrt{2} + \sqrt{\log_2 n}.$$

Hint. We already know that for some fixed l, $k_1 + k_2 + \cdots + k_l = n$. Naturally, we think of applying Cauchy inequality on the left hand side:

$$\sum_{i=1}^{l}\sqrt{\frac{k_i}{2^i}} \leqslant \sqrt{\sum_{i=1}^{l}k_i \cdot \sum_{i=1}^{l}\frac{1}{2^i}} \leqslant \sqrt{n \cdot 1} = \sqrt{n}. \qquad ①$$

Using the AM-GM inequality, however, leads to an even weaker result:

$$\sum_{i=1}^{l}\sqrt{\frac{k_i}{2^i}} \leqslant \sum_{i=1}^{l}\left(k_i + \frac{1}{2^i} \cdot \frac{1}{2}\right) = \frac{1}{2}n + \frac{1}{2}.$$

How can we use the fact that $\frac{1}{2^i} \leqslant a_j < \frac{1}{2^{i-1}}$? Note that if we shrink each $a_j \in \left[\frac{1}{2^i}, \frac{1}{2^{i-1}}\right]$ to $\frac{1}{2^i}$, we have

$$\sum_{i=1}^{l}k_i \cdot \frac{1}{2^i} \leqslant 1.$$

Hence from Cauchy inequality again we obtain

$$\sum_{i=1}^{l}\sqrt{\frac{k_i}{2^i}} \leqslant \sqrt{\sum_{i=1}^{l}1 \cdot \sum_{i=1}^{l}\frac{k_i}{2^i}} \leqslant \sqrt{l} \leqslant \sqrt{n}, \qquad ②$$

which does not seem to be an improvement after all. The culprits here involve $\sum_{i=1}^{l}\frac{1}{2^i}$ in ① and $\sum_{i=1}^{l}1$ in ②, and the remedy is to truncate the sums.

Proof.

$$\sum_{i=1}^{l}\sqrt{\frac{k_i}{2^i}} = \sum_{i\leqslant t}^{l}\sqrt{\frac{k_i}{2^i}} + \sum_{i>t}^{l}\sqrt{\frac{k_i}{2^i}},$$

where

$$\sum_{i\leqslant t}^{l}\sqrt{\frac{k_i}{2^i}} \leqslant \sqrt{\sum_{i=1}^{t}1 \cdot \sum_{i=1}^{t}\frac{k_i}{2^i}} \leqslant \sqrt{t},$$

and

$$\sum_{i>t}^{l}\sqrt{\frac{k_i}{2^i}} \leqslant \sqrt{\sum_{i=t+1}^{l}k_i \cdot \sum_{i=t+1}^{l}\frac{1}{2^i}} \leqslant \sqrt{n \cdot \frac{1}{2^t}}.$$

Let $t=\lfloor \log_2 n \rfloor$, we thus have

$$\text{LHS} \leqslant \sqrt{t}+\frac{\sqrt{n}}{2^{t/2}} \leqslant \sqrt{2}+\sqrt{\log_2 n}.$$

9.2 Enumeration Method

When all other methods have failed to make progress, enumeration method may just save the day. Start with situations that are easier to prove, but be careful not to miss any possible cases!

Eg. 4 Assume integer $n \geqslant 2$, $x_1, x_2, \cdots, x_n \in [0, 1]$. Prove there exists some i, $1 \leqslant i \leqslant n-1$ such that the following inequality holds:

$$x_i(1-x_{i+1}) \geqslant \frac{1}{4} x_1(1-x_n).$$

Proof. Let $m=\min\{x_1, x_2, \cdots, x_n\}$, and assume $x_r=m$, $0 \leqslant m \leqslant 1$. We consider the following two cases:

(1) If $x_2 \leqslant \frac{1}{2}(m+1)$, set $i=1$, we have

$$\begin{aligned} x_1(1-x_2) &\geqslant x_1\left(1-\frac{m+1}{2}\right)=\frac{1}{2} x_1(1-m) \\ &\geqslant \frac{1}{2} x_1(1-x_n) \text{ (since } m \leqslant x_n \leqslant 1) \\ &\geqslant \frac{1}{4} x_1(1-x_n). \end{aligned}$$

(2) If $x_2>\frac{1}{2}(m+1)$, we consider two possibilities:

(i) $x_1=m$, $x_2>\frac{1}{2}(1+m)$, $\cdots$, $x_n>\frac{1}{2}(1+m)$. Assume x_k is the smallest among $x_2, x_3, \cdots, x_n$ and set $i=k-1$, we have

$$x_{k-1}(1-x_k) \geqslant x_1(1-x_n) \geqslant \frac{1}{4} x_1(1-x_n),$$

where we have used the fact that $x_{k-1} \geqslant x_1$ and $1-x_k \geqslant 1-x_n$.

(ii) There exists some t, $3 \leqslant t \leqslant n$, such that $x_t = m \leqslant \frac{1}{2}(1+m)$. As a result, we can find an integer i, $2 \leqslant i \leqslant n-1$, satisfying

$$x_i > \frac{1}{2}(1+m), \; x_{i+1} \leqslant \frac{1}{2}(1+m).$$

For such i, we also have

$$x_i(1-x_{i+1}) > \frac{1}{2}(1+m)\left(1-\frac{m+1}{2}\right) = \frac{1}{2}(1+m) \cdot \frac{1}{2}(1-m)$$
$$= \frac{1}{4}(1-m^2) \geqslant \frac{1}{4}(1-m) \geqslant \frac{1}{4}x_1(1-x_n).$$

Eg. 5 Assume x_1, x_2, x_3, x_4, y_1 and y_2 satisfy

$$y_2 \geqslant y_1 \geqslant x_4 \geqslant x_3 \geqslant x_2 \geqslant x_1 \geqslant 2,$$
$$x_1 + x_2 + x_3 + x_4 \geqslant y_1 + y_2.$$

Prove that $x_1x_2x_3x_4 \geqslant y_1y_2$.

Proof. Keeping $y_1 + y_2$ the same while increasing y_1 to $\frac{y_1+y_2}{2}$, we find that y_1y_2 increases during the process. As a result, we only need to prove the case when $y_1 = y_2 = y$. In other words, assume $y \geqslant x_4 \geqslant x_3 \geqslant x_2 \geqslant x_1 \geqslant 2$ and $x_1 + x_2 + x_3 + x_4 \geqslant 2y$, we need to prove that

$$x_1x_2x_3x_4 \geqslant y^2.$$

Next, we discuss three possible cases:

(1) $y \geqslant 4$, then clearly $x_1x_2x_3x_4 \geqslant 2^4 \geqslant y^2$.

(2) $4 < y \leqslant 6$. Keep $x_1 + x_4$ and $x_2 + x_3$ the same, and adjust x_1, x_2 to 2.

Then keep $x_3 + x_4$ the same, and adjust x_3 to 2. Since $x_1x_2x_3x_4$ decreases during the adjustment, it suffices to show that

$$2^3 \cdot x_4 \geqslant y^2.$$

Because $2^3 \cdot x_4 \geqslant 8(2y-6)$, we only need to show that $8(2y-6) \geqslant y^2$, or equivalently, $(y-4)(y-12) \leqslant 0$. This holds obviously as $4 < y \leqslant 6$.

(3) $y > 6$. Keep the sums $x_1 + x_4$ and $x_2 + x_3$ constant and make adjustments according to the following three cases:

(i) If both of these two sums $\geqslant y + 2$, then adjust x_4 and x_3 to y. The conclusion obviously holds.

(ii) If only one of the sums $\geqslant y + 2$. Without loss of generality assume $x_1 + x_4 \geqslant y + 2$. Then we can adjust x_4 to y and x_2 to 2. Next, we keep $x_1 + x_3$ the same and adjust x_1 to 2, then

$$x_1 x_2 x_3 x_4 \geqslant 2^2 y(y-4) > y^2.$$

(iii) If both sums $< y + 2$, we adjust x_1 and x_2 to 2, then keep $x_3 + x_4$ the same and adjust the larger one between x_3 and x_4 to y. The refore, $x_1 x_2 x_3 x_4 \geqslant 2^2 y(y-4) > y^2$ so the conclusion follows.

Eg. 6 Let $a_1, a_2, \cdots, a_n (n \geqslant 3)$ be real numbers, prove that

$$\sum_{i=1}^{n} a_i^2 - \sum_{i=1}^{n} a_i a_{i+1} \leqslant \left\lfloor \frac{n}{2} \right\rfloor (M-m)^2,$$

where $a_{n+1} = a_1$, $M = \max\limits_{1 \leqslant i \leqslant n} a_i$, $m = \min\limits_{1 \leqslant i \leqslant n} a_i$, and $\lfloor x \rfloor$ represents the largest integer not greater than x.

(China Mathematical Olympiad, 2011)

Proof. If $n = 2k$ (k a positive integer), we have

$$2\left(\sum_{i=1}^{n} a_i^2 - \sum_{i=1}^{n} a_i a_{i+1}\right) = \sum_{i=1}^{n} (a_i - a_{i+1})^2 \leqslant n \cdot (M-m)^2.$$

Thus

$$\sum_{i=1}^{n} a_i^2 - \sum_{i=1}^{n} a_i a_{i+1} \leqslant \frac{n}{2} \cdot (M-m)^2 = \left\lfloor \frac{n}{2} \right\rfloor (M-m)^2.$$

If $n = 2k + 1$ (k a positive integer), then there must exist three consecutive terms (the neighbors of each term a_i is determined by placing $a_1, \cdots, a_n$ on a circle) that are non-increasing or non-decreasing. This is because $\prod\limits_{i=1}^{2k+1} (a_i - a_{i-1})(a_{i+1} - a_i) = \prod\limits_{i=1}^{2k+1} (a_i - a_{i-1})^2 \geqslant 0$, so it is impossible for each i that $a_i - a_{i-1}$ and $a_{i+1} - a_i$ have different signs. Without loss of generality, assume a_1, a_2 and a_3 are non-increasing or non-decreasing. Thus,

$$(a_1 - a_2)^2 + (a_2 - a_3)^2 \leqslant (a_1 - a_3)^2.$$

In consequence,

$$2\left(\sum_{i=1}^{n} a_i^2 - \sum_{i=1}^{n} a_i a_{i+1}\right) = \sum_{i=1}^{n} (a_i - a_{i+1})^2 \leqslant (a_1 - a_3)^2 + \sum_{i=3}^{n} (a_i - a_{i+1})^2,$$

which reduces the problem to the case of $2k$ terms. Therefore,

$$2\left(\sum_{i=1}^{n} a_i^2 - \sum_{i=1}^{n} a_i a_{i+1}\right) \leqslant (a_1 - a_3)^2 + \sum_{i=3}^{n} (a_i - a_{i+1})^2 \leqslant 2k(M - m)^2.$$

In other words,

$$\left(\sum_{i=1}^{n} a_i^2 - \sum_{i=1}^{n} a_i a_{i+1}\right) \leqslant k(M - m)^2 = \left\lfloor \frac{n}{2} \right\rfloor (M - m)^2.$$

9.3 Add Ordering Conditions

Introducing ordering in the proof of an inequality can give us useful conditions to rely on.

Eg. 7 (1) If x, y, z are positive integers that are not all the same, find the minimum value of $(x + y + z)^3 - 27xyz$.

(2) If x, y, z are positive integers different from each other, find the minimum value of $(x + y + z)^3 - 27xyz$.

Soln.

$$\begin{aligned}
&(x + y + z)^3 - 27xyz \\
&= x^3 + y^3 + z^3 + 3(x^2y + y^2z + z^2x) + 3(xy^2 + yz^2 + zx^2) \\
&\quad + 6xyz - 27xyz \\
&= (x + y + z)(x^2 + y^2 + z^2 - xy - yz - zx) \\
&\quad + 3(x^2y + y^2z + z^2x + xy^2 + yz^2 + zx^2 - 6xyz) \\
&= \frac{x + y + z}{2}[(x - y)^2 + (y - z)^2 + (z - x)^2] \\
&\quad + 3[x(y - z)^2 + y(z - x)^2 + z(x - y)^2].
\end{aligned}$$

(1) Without loss of generality, assume $x \geqslant y \geqslant z$, so $z \geqslant 1$, $y \geqslant$

$1, x \geqslant 2$.

Hence,

$$\begin{aligned}
&(x+y+z)^3-27xyz \\
\geqslant\ &2[(x-y)^2+(y-z)^2+(z-x)^2] \\
&+3[(y-z)^2+(z-x)^2+(x-y)^2] \\
=\ &5[(x-y)^2+(y-z)^2+(z-x)^2] \geqslant 10.
\end{aligned}$$

Thus $(x+y+z)^3-27xyz \geqslant 10$, and the equal sign is achieved when $(x, y, z)=(2, 1, 1)$.

(2) Without loss of generality, assume $x>y>z$, so $z \geqslant 1$, $y \geqslant 2$, $x \geqslant 3$.

Hence,

$$\begin{aligned}
&(x+y+z)^3-27xyz \\
\geqslant\ &3[(x-y)^2+(y-z)^2+(z-x)^2] \\
&+3[3(y-z)^2+2(z-x)^2+(x-y)^2] \\
\geqslant\ &3 \cdot (1^2+1^2+2^2)+3 \cdot (3 \cdot 1^2+2 \cdot 2^2+1^2)=54.
\end{aligned}$$

Therefore $(x+y+z)^3-27xyz \geqslant 54$, and the equal sign is achieved when $(x, y, z)=(3, 2, 1)$.

[Comment] Part (1) is equivalent to the following problem: For integers a, b, c that are not all the same, we have

$$\frac{a+b+c}{3} \geqslant \sqrt[3]{abc+\frac{10}{27}}. \qquad ①$$

Of course, we can use the following alternative method to prove ①: assume $1 \leqslant a \leqslant b \leqslant c$, $b=a+x$, $c=a+y$. Then $x, y \geqslant 0$, and x, y are not both $0 (x \leqslant y)$. ① is equivalent to

$$9a(x^2-xy+y^2)+(x+y)^3 \geqslant 10.$$

Since $a \geqslant 1$, $x^2-xy+y^2 \geqslant 1$, $x+y \geqslant 1$, the above is obviously true, and equality is achieved when $(a, x, y)=(1, 0, 1)$.

Eg. 8 Assume $a_1, a_2, \cdots$ to be an infinite sequence of real numbers. There exists a real number c such that $0 \leqslant a_i \leqslant c$ or all i. In addition, for all $i \neq j$, $|a_i-a_j| \geqslant \frac{1}{i+j}$. Prove that $c \geqslant 1$.

Proof. For a fixed $n \geqslant 2$, order the first n terms of the series as

$$0 \leqslant a_{\sigma(1)} < a_{\sigma(2)} < \cdots < a_{\sigma(n)} \leqslant c,$$

where $\sigma(1), \sigma(2), \cdots, \sigma(n)$ is some permutation of $1, 2, \cdots, n$. Thus,

$$\begin{aligned} c &\geqslant a_{\sigma(n)} - a_{\sigma(1)} \\ &= (a_{\sigma(n)} - a_{\sigma(n-1)}) + (a_{\sigma(n-1)} - a_{\sigma(n-2)}) + \cdots + (a_{\sigma(2)} - a_{\sigma(1)}) \\ &\geqslant \frac{1}{\sigma(n) + \sigma(n-1)} + \frac{1}{\sigma(n-1) + \sigma(n-2)} + \cdots + \frac{1}{\sigma(2) + \sigma(1)}. \end{aligned} \quad ①$$

From Cauchy Inequality, we have

$$\sum_{i=1}^{n-1} \frac{1}{\sigma(n+1-i) + \sigma(n-i)} \sum_{i=1}^{n-1} (\sigma(n+1-i) + \sigma(n-i)) \geqslant (n-1)^2.$$

Thus

$$\begin{aligned} &\frac{1}{\sigma(n) + \sigma(n-1)} + \frac{1}{\sigma(n-1) + \sigma(n-2)} + \cdots + \frac{1}{\sigma(2) + \sigma(1)} \\ \geqslant &\frac{(n-1)^2}{2(\sigma(1) + \cdots + \sigma(n)) - \sigma(1) - \sigma(n)} \\ = &\frac{(n-1)^2}{n(n+1) - \sigma(1) - \sigma(n)} \\ \geqslant &\frac{(n-1)^2}{n^2 + n - 3} \geqslant \frac{n-1}{n+3}. \end{aligned}$$

From ①, $c \geqslant 1 - \frac{4}{n+3}$ for all $n \geqslant 2$. Therefore $c \geqslant 1$.

9.4 Treatment of Asymmetry in Inequality

Asymmetry in an inequality occurs when the roles of variables are rather different. In such case, it is important to investigate when the equality holds, and we often need to work out ways to put each variable on an equal footing again.

Eg. 9 Assume that $a \leqslant b < c$ are the three sides of a right triangle. Find the largest constant M such that $\frac{1}{a} + \frac{1}{b} + \frac{1}{c} \geqslant \frac{M}{a+b+c}$ always holds.

Hint. For the familiar isosceles right triangle, a and b are the same, we guess this may well give the largest M and we want to separate a and b from c in our derivation.

Soln. When $a = b = \frac{\sqrt{2}}{2}c$, $M \leqslant 2+3\sqrt{2}$, we next show that given the conditions,

$$\frac{1}{a} + \frac{1}{b} + \frac{1}{c} \geqslant \frac{2+3\sqrt{2}}{a+b+c},$$

or

$$a^2(b+c) + b^2(c+a) + c^2(a+b) \geqslant (2+3\sqrt{2})abc$$

always holds. In fact, the left hand side of the above is

$$\begin{aligned} & c(a^2+b^2) + \left(a^2b + b \cdot \frac{c^2}{2}\right) + \left(ab^2 + a \cdot \frac{c^2}{2}\right) + \frac{1}{2}c^2(a+b) \\ \geqslant\ & 2abc + \sqrt{2}abc + \sqrt{2}abc + \frac{1}{2}c \cdot \sqrt{a^2+b^2} \cdot (a+b) \\ \geqslant\ & 2abc + 2\sqrt{2}abc + \frac{1}{2}c \cdot \sqrt{2ab} \cdot 2\sqrt{ab} \\ =\ & (2+3\sqrt{2})abc. \end{aligned}$$

Therefore $M_{\max} = 2+3\sqrt{2}$.

Eg. 10 Find the maximum value of real number a such that for any real numbers x_1, x_2, x_3, x_4, x_5, the inequality

$$x_1^2 + x_2^2 + x_3^2 + x_4^2 + x_5^2 \geqslant a(x_1x_2 + x_2x_3 + x_3x_4 + x_4x_5)$$

always holds.

(Mathematics Competition of Senior High School of Shanghai, 2010)

Soln. Set $x_1 = 1$, $x_2 = \sqrt{3}$, $x_3 = 2$, $x_4 = \sqrt{3}$, $x_5 = 1$, we have $a \leqslant \frac{2}{\sqrt{3}}$. On the other hand,

$$x_1^2 + x_2^2 + x_3^2 + x_4^2 + x_5^2 = \left(x_1^2 + \frac{x_2^2}{3}\right) + \left(\frac{2x_2^2}{3} + \frac{x_3^2}{2}\right) + \left(\frac{x_3^2}{2} + \frac{2x_4^2}{3}\right) + \left(\frac{x_4^2}{3} + x_5^2\right)$$

$$\geqslant \frac{2}{\sqrt{3}}x_1x_2 + \frac{2}{\sqrt{3}}x_2x_3 + \frac{2}{\sqrt{3}}x_3x_4 + \frac{2}{\sqrt{3}}x_4x_5.$$

Therefore, $a_{\max} = \frac{2}{\sqrt{3}} = \frac{2\sqrt{3}}{3}$.

Eg. 11 Assume that a, b, c are positive real numbers with $a+b+c=10$, and $a \leqslant 2b$, $b \leqslant 2c$, $c \leqslant 2a$, find the minimum value of abc.

Soln. Denote $x = 2b - a$, $y = 2c - b$, $z = 2a - c$, the original constraints are equivalent to

$$x + y + z = 10, \ x \geqslant 0, \ y \geqslant 0, \ z \geqslant 0.$$

Also, $a = \frac{x+2y+4z}{7}$, $b = \frac{y+2z+4x}{7}$, $c = \frac{z+2x+4y}{7}$, thus

$$abc = \frac{1}{343} \cdot [(x+2y+4z)(y+2z+4x)(z+2x+4y)].$$

Note that

$$\begin{aligned}
&(x+2y+4z)(y+2z+4x)(z+2x+4y)\\
&= (10+y+3z)(10+z+3x)(10+x+3y)\\
&= 1000 + 400(x+y+z) + 130(xy+yz+zx) + 30(x^2+y^2+z^2)\\
&\quad + (3y^2z + 3x^2y + 9xy^2 + 3xz^2 + 9yz^2 + 9x^2z + 28xyz)\\
&= 1000 + 4000 + 30(x+y+z)^2 + [70(xy+yz+zx) +\\
&\quad 3(y^2z + x^2y + xz^2) + 9(xy^2 + yz^2 + x^2z) + 28xyz]\\
&\geqslant 8000.
\end{aligned}$$

The equality holds when $x = y = 0$, $z = 10$, or when $a = \frac{40}{7}$, $b = \frac{20}{7}$, $c = \frac{10}{7}$.

Thus $abc_{\min} = \frac{8000}{343}$.

Eg. 12 Let a, b, c, d be positive real numbers and $abcd = 1$. Let $T = a(b+c+d) + b(c+d) + cd$.

(1) Find the minimum value of $a^2 + b^2 + T$;

(2) Find the minimum value of $a^2 + b^2 + c^2 + T$.

Hint. For (1), we should consider a and b jointly, and c and d jointly. For (2), we should treat a, b, c indifferently.

Soln. (1) From the hint, we rewrite T accordingly:

$$\begin{aligned} a^2+b^2+T &= a^2+b^2+(a+b)(c+d)+ab+cd \\ &\geqslant 2ab+2\sqrt{ab}\cdot 2\sqrt{cd}+ab+cd \\ &= 4+3ab+cd \geqslant 4+2\cdot\sqrt{3abcd} \\ &= 4+2\sqrt{3}. \end{aligned}$$

When $a=b=\left(\frac{1}{3}\right)^{1/4}$, $c=d=3^{1/4}$, the equality holds.

(2) Again, we rewrite T so that a, b, c are treated similarly:

$$\begin{aligned} a^2+b^2+c^2+T &= a^2+b^2+c^2+(a+b+c)d+ab+bc+ca \\ &\geqslant 3\sqrt[3]{a^2b^2c^2}+3\sqrt[3]{abc}\cdot d+3\cdot\sqrt[3]{(abc)^2} \\ &= 3\cdot[2\sqrt[3]{(abc)^2}+\sqrt[3]{abc}\cdot d] \\ &\geqslant 3\cdot 2\cdot\sqrt{2\sqrt[3]{(abc)^2}\cdot\sqrt[3]{abc}\cdot d} = 6\sqrt{2}. \end{aligned}$$

When $a=b=c=\left(\frac{1}{2}\right)^{\frac{1}{4}}$, $d=2^{\frac{3}{4}}$, the equality holds.

Exercises 9

1. Assume $n\geqslant a_1>\cdots>a_k\geqslant 1$ satisfying: for any i, j, $[a_i, a_j]\leqslant n$. (Note: $[x, y]$ represents the least common multiple of integers x and y.) Show that $k\leqslant 2\sqrt{n}+1$.

2. For any $x\in\mathbf{R}$ and $n\in\mathbf{N}_+$, show that

(1) $\left|\sum\limits_{k=1}^{n}\frac{\sin kx}{k}\right|\leqslant 2\sqrt{\pi}$.

(2) $\sum\limits_{k=1}^{n}\frac{|\sin kx|}{k}\geqslant|\sin nx|$.

3. Let x_1, x_2, $\cdots$, x_n $(n\geqslant 2)$ be positive numbers. Prove that

$$\frac{x_1^2}{x_1^2+x_2x_3}+\frac{x_2^2}{x_2^2+x_3x_4}+\cdots+\frac{x_{n-1}^2}{x_{n-1}^2+x_nx_1}+\frac{x_n^2}{x_n^2+x_1x_2}\leqslant n-1.$$

4. Let $x_1, x_2, \cdots, x_n$ $(n \geqslant 3)$ be non-negative real numbers such that

$$\sum_{i=1}^{n} x_i^2 + \sum_{1 \leqslant i < j \leqslant n} (x_i x_j)^2 = \frac{n(n+1)}{2}.$$

(1) Find the maximum value of $\sum_{i=1}^{n} x_i$.

(2) Find all positive integers n, such that $\sum_{i=1}^{n} x_i \geqslant \sqrt{\frac{n(n+1)}{2}}$.

5. Assume that $a \geqslant b \geqslant c > 0$, and $a + b + c = 3$. Prove that

$$\frac{a}{c} + \frac{b}{a} + \frac{c}{b} \geqslant 3 + |(a-1)(b-1)(c-1)|.$$

6. Let a, b, c be positive real numbers and $abc \leqslant 1$. Prove that

$$\frac{a}{c} + \frac{b}{a} + \frac{c}{b} \geqslant a + b + c + |(a-1)(b-1)(c-1)|.$$

7. Place six circles of radius 1, 2, 3, 4, 5 and 6 on the x-axis so that the x-axis is a common tangent line to all six circles and the adjacent circles are externally tangent to each other. Find the minimum and maximum length of the common tangent line segment between the leftmost and rightmost circles.

8. Assume five positive numbers satisfy the following conditions:

(1) One of them is $\frac{1}{2}$;

(2) Pick any two numbers from these five numbers, there exist one number from the remaining three such that the sum of this number and the previously picked two numbers is exactly 1.

Find these five numbers.

9. Let x, y, z be non-negative real numbers and $x + y + z = 1$. Find the maximum value of $x^2y^2 + y^2z^2 + z^2x^2 + x^2y^2z^2$.

10. Let x, y, z be non-negative real numbers and $x + y + z = 1$. Find the minimum and maximum value of $x^2y + y^2z + z^2x$.

11. Assume a, b, c are the three sides of a triangle. Show that

$$\left| \frac{(a-b)(b-c)(c-a)}{(a+b)(b+c)(c+a)} \right| < \frac{1}{22}.$$

12. Find the largest constant k, such that $\frac{kabc}{a+b+c} \leqslant (a+b)^2 + (a+b+4c)^2$ holds for any positive real numbers a, b, c.

13. (1) Prove that for any real numbers p, q,

$$p^2 + q^2 + 1 > p(q+1).$$

(2) Find the largest real number b such that for any real numbers p, q, $p^2 + q^2 + 1 > bp(q+1)$.

(3) Find the largest real number c such that for any integers p, q,

$$p^2 + q^2 + 1 > cp(q+1).$$

14. (1) Assume a, b, c are the three sides of a triangle, $n \geqslant 2$ is an integer. Show that

$$\frac{\sqrt[n]{a^n + b^n} + \sqrt[n]{b^n + c^n} + \sqrt[n]{c^n + a^n}}{a+b+c} < 1 + \frac{\sqrt[n]{2}}{2}.$$

(2) Assume a, b, c are the three sides of a triangle. Find the smallest positive real number k, such that the following inequality always holds:

$$\frac{\sqrt[3]{a^3 + b^3} + \sqrt[3]{b^3 + c^3} + \sqrt[3]{c^3 + a^3}}{(\sqrt{a} + \sqrt{b} + \sqrt{c})^2} < k.$$

Detailed Solutions to Exercises

Exercise 1

1. Denote $a = x + y + z$, $b = xy + yz + zx$, we have $x^2 + y^2 + z^2 = a^2 - 2b$. Thus LHS - RHS $= 2b^2(a^2 - 3b) = b^2 \cdot [(x - y)^2 + (y - z)^2 + (z - x)^2] \geqslant 0$.

2. It suffices to show that $\left(1 + \frac{1}{n}\right)^n < \left(1 + \frac{1}{n+1}\right)^{n+1}$.

When $n = 1$, the above obviously holds. When $n \geqslant 2$, according to the binomial theorem,

$$\left(1 + \frac{1}{n}\right)^n = \sum_{k=2}^{n} \frac{1}{k!} \cdot \left(1 - \frac{1}{n}\right)\left(1 - \frac{2}{n}\right)\cdots\left(1 - \frac{k-1}{n}\right) + 2.$$

$$\begin{aligned}\left(1 + \frac{1}{n+1}\right)^{n+1} &= \sum_{k=2}^{n+1} \frac{1}{k!} \cdot \left(1 - \frac{1}{n+1}\right)\left(1 - \frac{2}{n+1}\right)\cdots\left(1 - \frac{k-1}{n+1}\right) + 2 \\ &= \sum_{k=2}^{n} \frac{1}{k!} \cdot \left(1 - \frac{1}{n+1}\right)\left(1 - \frac{2}{n+1}\right)\cdots \\ &\quad \left(1 - \frac{k-1}{n+1}\right) + \left(\frac{1}{n+1}\right)^{n+1} + 2.\end{aligned}$$

So clearly $\left(1 + \frac{1}{n}\right)^n < \left(1 + \frac{1}{n+1}\right)^{n+1}$.

3. Since $A_n > 0$, $B_n > 0$, it suffices to show that $A_n B_{n-1} - A_{n-1} B_n > 0$. Note that

$$\begin{aligned}A_n B_{n-1} - A_{n-1} B_n &= (x_n a^n + A_{n-1}) B_{n-1} - A_{n-1}(x_n b^n + B_{n-1}) \\ &= x_n (a^n B_{n-1} - b^n A_{n-1}) \\ &= x_n [x_{n-1}(a^n b^{n-1} - a^{n-1} b^n) + x_{n-2}(a^n b^{n-2} - a^{n-2} b^n) \\ &\quad + \cdots + x_0 (a^n - b^n)].\end{aligned}$$

Since $a > b$, when $1 \leqslant k \leqslant n$, we have $a^n b^{n-k} > a^{n-k} b^n$. Also $x_i \geqslant 0 (i = 0, 1, \cdots, n-2)$, x_{n-1}, $x_n \geqslant 0$, thus $A_n B_{n-1} - A_{n-1} B_n > 0$.

4. Using $a^2 + b^2 + c^2 \geqslant ab + bc + ca$ $(a, b, c \in \mathbf{R})$, we have

$$\begin{aligned} a^8 + b^8 + c^8 &\geqslant a^4 b^4 + b^4 c^4 + c^4 a^4 \geqslant a^2 b^4 c^2 + b^2 c^4 a^2 + c^2 a^4 b^2 \\ &= a^2 b^2 c^2 (a^2 + b^2 + c^2) \geqslant a^2 b^2 c^2 (ab + bc + ca). \end{aligned}$$

Divide each side by $a^3 b^3 c^3$ and we have proved the original inequality.

5. Note that

$$\begin{aligned} & a_1^2 + a_2^2 + \cdots + a_{100}^2 \leqslant (100 - a_2)^2 + a_2^2 + a_3^2 + \cdots + a_{100}^2 \\ \leqslant\ & 100^2 - (a_1 + a_2 + \cdots + a_{100}) a_2 + 2a_2^2 + a_3^2 + \cdots + a_{100}^2 \\ =\ & 100^2 - a_2(a_1 - a_2) - a_3(a_2 - a_3) - \cdots - a_{100}(a_2 - a_{100}) \\ \leqslant\ & 100^2. \end{aligned}$$

The equality holds when $a_1 = 100$, $a_i = 0 (i > 1)$ or when $a_1 = a_2 = a_3 = a_4 = 50$, $a_j = 0 (j > 4)$.

6. Let us prove a Lemma first:

Lemma: Assume $x_1, x_2, \cdots, x_n$ are n real numbers greater than 1, $A = \sqrt[n]{x_1 x_2 \cdots x_n}$, then

$$\prod_{i=1}^{n} \frac{(x_i + 1)}{(x_i - 1)} \geqslant \left(\frac{A+1}{A-1}\right)^n.$$

Proof: WLOG assume $x_1 \leqslant x_2 \leqslant \cdots \leqslant x_n$, thus $x_1 \leqslant A \leqslant x_n$. Via reduction to common denominator, it is easy to show that

$$\frac{(x_1 + 1)(x_n + 1)}{(x_1 - 1)(x_n - 1)} \geqslant \left(\frac{A+1}{A-1}\right) \cdot \left(\frac{\dfrac{x_1 x_n}{A} + 1}{\dfrac{x_1 x_n}{A} - 1}\right).$$

Therefore,

$$\prod_{i=1}^{n} \left(\frac{x_i + 1}{x_i - 1}\right) \geqslant \prod_{i=2}^{n} \left(\frac{x_i + 1}{x_i - 1}\right) \left(\frac{\dfrac{x_1 x_n}{A} + 1}{\dfrac{x_1 x_n}{A} - 1}\right) \left(\frac{A+1}{A-1}\right).$$

Consider the remaining $n-1$ real numbers $x_2, x_3, \cdots, x_{n-1}$ and x_1x_n, whose geometric average is still A. The maximum of these $n-1$ numbers is no smaller than A and the minimum is no larger than A. Using the same methods as before, after repeating $n-1$ times we will have $\prod_{i=1}^{n}\left(\frac{x_i+1}{x_i-1}\right)\geqslant\left(\frac{A+1}{A-1}\right)^n$.

Now let's prove the original statement. Let $x_i = r_is_it_iu_iv_i(1\leqslant i\leqslant n)$, from the lemma we have

$$\prod_{i=1}^{n}\frac{(r_is_it_iu_iv_i+1)}{(r_is_it_iu_iv_i-1)}\geqslant\left(\frac{B+1}{B-1}\right)^n,$$

where $B=\sqrt[n]{\prod_{i=1}^{n}(r_is_it_iu_iv_i)}$. Hence it suffices to show that

$$\frac{B+1}{B-1}\geqslant\frac{RSTUV+1}{RSTUV-1}.$$

Note that

$$\begin{aligned}RSTUV &= \frac{1}{n}\cdot\left(\sum_{i=1}^{n}r_i\right)\cdot\frac{1}{n}\cdot\left(\sum_{i=1}^{n}s_i\right)\cdot\frac{1}{n}\cdot\left(\sum_{i=1}^{n}t_i\right)\cdot\frac{1}{n}\cdot\\ &\quad\left(\sum_{i=1}^{n}u_i\right)\cdot\frac{1}{n}\cdot\left(\sum_{i=1}^{n}v_i\right)\\ &\geqslant\sqrt[n]{\prod_{i=1}^{n}r_i}\cdot\sqrt[n]{\prod_{i=1}^{n}s_i}\cdot\sqrt[n]{\prod_{i=1}^{n}t_i}\cdot\sqrt[n]{\prod_{i=1}^{n}u_i}\cdot\sqrt[n]{\prod_{i=1}^{n}v_i}=B,\end{aligned}$$

Thus $(B+1)(RSTUV-1)-(B-1)(RSTUV+1)=2(RSTUV-B)\geqslant 0$.

Therefore the original inequality holds.

Comment: We can also use Jensen's Inequality.

First, it's easy to show that, for any $a, b>1$,

$$\left(\frac{a+1}{a-1}\right)\cdot\left(\frac{b+1}{b-1}\right)\geqslant\left[\frac{\sqrt{ab}+1}{\sqrt{ab}-1}\right]^2.$$

Hence on the interval $(0, +\infty)$, the function $y=\ln\left(\frac{e^x+1}{e^x-1}\right)$ is a convex. This implies

$$\prod_{i=1}^{n}\left(\frac{r_is_it_iu_iv_i+1}{r_is_it_iu_iv_i-1}\right)\geqslant\frac{\sqrt[n]{\prod_{i=1}^{n}(r_is_it_iu_iv_i)}+1}{\sqrt[n]{\prod_{i=1}^{n}(r_is_it_iu_iv_i)}-1}\geqslant\left(\frac{RSTUV+1}{RSTUV-1}\right)^n.$$

7. Denote $T=x_1x_2\cdots x_k=x_1+x_2+\cdots+x_k$. From the AM-GM Inequality, $T/k\geqslant T^{1/k}$, $x_1^{n-1}+x_2^{n-1}+\cdots+x_k^{n-1}\geqslant k\cdot T^{(n-1)/k}$. Therefore it suffices to show that $T^{(n-1)/k}\geqslant n$. Since $T/k\geqslant T^{1/k}$, $T^{(n-1)/k}\geqslant k^{(n-1)/(k-1)}$, we only need to show that $k^{(n-1)/(k-1)}\geqslant n$, or $k\geqslant n^{(k-1)/(n-1)}$. In fact,

$$k=\frac{(k-1)n+(n-k)\cdot 1}{n-1}\geqslant\sqrt[n-1]{n^{k-1}\cdot 1^{n-k}}=n^{(k-1)/(n-1)},$$

so the conclusion holds.

8. If we can show that $\frac{27}{64}(a+b)^2(b+c)^2(c+a)^2\geqslant(ab+bc+ca)^2$, then the conclusion follows. Denote $S_1=a+b+c$, $S_2=ab+bc+ca$, $S_3=abc$. We need to show that $27(S_1S_2-S_3)^2\geqslant 64S_2^3$. Consider the following two cases:

(1) If a, b, c are all non-negative, then

$$27(S_1S_2-S_3)^2\geqslant 27\left(S_1S_2-\frac{1}{9}S_1S_2\right)^2=\frac{64}{3}S_1^2S_2^2\geqslant 64S_2^3.$$

(2) If at least one of a, b, c is negative. By symmetry, WLOG assume that $a<0$, $b\geqslant 0$, $c\geqslant 0$. Assume $S_2>0$ (otherwise the statement is obviously true). Then $S_1>0$, $S_3<0$, thus

$$\begin{aligned}&27(S_1S_2-S_3)^2-64S_2^3>27S_1^2S_2^2-64S_2^3=S_2^2\cdot(27S_1^2-64S_2)\\&=S_2^2\cdot[27a^2+22(b^2+c^2)+5(b-c)^2-10a(b+c)]>0.\end{aligned}$$

The equality holds if and only if $a=b=c$.

9. Consider

$$\begin{aligned}S&=\sum_{\varepsilon_1,\varepsilon_2,\cdots,\varepsilon_n}|(a_1\varepsilon_1+a_2\varepsilon_2+\cdots+a_n\varepsilon_n)|^2\\&=\sum_{\varepsilon_1,\varepsilon_2,\cdots,\varepsilon_n}(a_1\varepsilon_1+a_2\varepsilon_2+\cdots+a_n\varepsilon_n)(\bar{a}_1\varepsilon_1+\bar{a}_2\varepsilon_2+\cdots+\bar{a}_n\varepsilon_n)\\&=\sum_{\varepsilon_1,\varepsilon_2,\cdots,\varepsilon_n}(|a_1|^2+|a_2|^2+\cdots+|a_n|^2+\sum_{i\neq j}a_i\bar{a}_j\varepsilon_i\varepsilon_j).\end{aligned}$$

WLOG assume that $|a_i|=1$, $i=1,2,\cdots,n$, thus

$$S=n\cdot 2^n+\sum_{\varepsilon_1,\varepsilon_2,\cdots,\varepsilon_n}\sum_{i\neq j}a_i\bar{a}_j\varepsilon_i\varepsilon_j=n\cdot 2^n+\sum_{i\neq j}a_i\bar{a}_j\sum_{\varepsilon_1,\varepsilon_2,\cdots,\varepsilon_n}\varepsilon_i\varepsilon_j$$
$$=n\cdot 2^n.$$

Since we require $n^{2/3}>c\cdot\frac{1}{2^n}\cdot\sum_{\varepsilon_1,\varepsilon_2,\cdots,\varepsilon_n}|a_1\varepsilon_1+a_2\varepsilon_2+\cdots+a_n\varepsilon_n|$, we want $c\cdot\sqrt{\frac{S}{2^n}}=c\cdot\sqrt{n}\leqslant n^{2/3}$ which implies $n\geqslant c^6$.

10. Let $\sqrt{u}=a\cdot\sqrt{\sin\theta}$, $\sqrt{v}=b\cdot\sqrt{\cos\theta}$, the original conditions are equivalent to $\frac{u^2}{a^2}+\frac{v^2}{b^4}=1$, and $u, v\geqslant 0$. From Cauchy Inequality, we have

$$y=\sqrt{u}+\sqrt{v}\leqslant\sqrt{\left(\frac{u}{a^{4/3}}+\frac{v}{b^{4/3}}\right)(a^{4/3}+b^{4/3})}$$
$$\leqslant\sqrt{\sqrt{\left(\frac{u^2}{a^2}+\frac{v^2}{b^4}\right)(a^{4/3}+b^{4/3})}\cdot(a^{4/3}+b^{4/3})}=(a^{4/3}+b^{4/3})^{3/4}.$$

It's easy to check when the equality holds.

11. Denote these n real numbers as $y_1, y_2, \cdots, y_n$. Let $x_i=y_i/2$, then $|x_i|\leqslant 1$, $\sum_{i=1}^{n}x_i^3=0$. We need to show that $\sum_{i=1}^{n}x_i\leqslant\frac{1}{3}n$.

We use the method of undetermined coefficient: assume $x_i\leqslant\frac{1}{3}+\lambda x_i^3$. Set $x=\cos\theta$, it's easy to see that $\lambda=\frac{4}{3}$ will work according to the triple angle formula, this concludes the proof of the original inequality.

12. Since $x_k\in[-1,1]$, $0\leqslant(1+x_k)(Bx_k-1)^4$, where B is an undetermined coefficient. Expanding this, we obtain

$$B^4x_k^5+(B^4-4B^3)x_k^4+B^2(6-4B)x_k^3+B(4B-6)x_k^2+$$
$$(1-4B)x_k+1\geqslant 0.$$

Let $B=\frac{3}{2}$, the last inequality becomes

$$\frac{81}{16}x_k^5 - \frac{135}{16}x_k^4 + \frac{15}{2}x_k^2 - 5x_k + 1 \geqslant 0.$$

Take the sum from $k = 1$ to n, we obtain

$$5\sum_{k=1}^{n} x_k \leqslant n - \frac{135}{16}\sum_{k=1}^{n} x_k^4 + \frac{15}{2}\sum_{k=1}^{n} x_k^2$$

$$= n + \frac{15}{2}\left(\sum_{k=1}^{n} x_k^2 - \frac{9}{8}\sum_{k=1}^{n} x_k^4\right)$$

$$\leqslant n + \frac{15}{2} \cdot \frac{2}{9}n = \frac{8}{3}n.$$

Hence $\sum_{k=1}^{n} x_k \leqslant \frac{8}{15}n$.

13. First consider the case when $a = 0$. We have $\sum_{k=1}^{n} x_k^2 = -2\sum_{1\leqslant i<j\leqslant n} x_i x_j \leqslant 2\sum_{i\neq j} |x_i x_j|$, thus

$$2\sum_{k=1}^{n} x_k^2 \leqslant \sum_{k=1}^{n} x_k^2 + 2\sum_{i\neq j} |x_i x_j| = \left(\sum_{k=1}^{n} |x_k|\right)^2.$$

When $a \neq 0$, let $y_k = x_k - a$, then the arithmetic mean of $y_1, y_2, \cdots, y_n$ is 0. Similar to the previous case $\sum_{k=1}^{n} y_k^2 \leqslant \frac{1}{2}\left(\sum_{k=1}^{n} |y_k|\right)^2$, therefore the original inequality holds.

14. WLOG assume $x + y + z = 1$, the original inequality is equivalent to

$$4x^3 + 4y^3 + 4z^3 - 4(x^2 + y^2 + z^2) + 1 \geqslant 3xyz.$$

Since $x^3 + y^3 + z^3 = 1 + 3xyz - 3(xy + yz + zx)$, $x^2 + y^2 + z^2 = 1 - 2(xy + yz + zx)$, the original inequality is equivalent to

$$xy + yz + zx - \frac{9}{4}xyz \leqslant \frac{1}{4}.$$

WLOG assume that $x \geqslant y \geqslant z \geqslant 0$, then $x + y \geqslant \frac{2}{3}$, $z \leqslant \frac{1}{3}$. Assume $x + y = \frac{2}{3} + \delta$, $z = \frac{1}{3} - \delta$, where $\delta \in \left[0, \frac{1}{3}\right]$. Therefore,

$$\begin{aligned}
xy+z(x+y)-\frac{9}{4}xyz &= xy+\left(\frac{1}{3}-\delta\right)\left(\frac{2}{3}+\delta\right)-\frac{9}{4}xy\left(\frac{1}{3}-\delta\right)\\
&= xy\left(\frac{1}{4}+\frac{9}{4}\delta\right)+\frac{2}{9}-\delta^2-\frac{1}{3}\delta\\
&\leqslant \left(\frac{1}{3}+\frac{1}{2}\delta\right)^2\cdot\left(\frac{1}{4}+\frac{9}{4}\delta\right)+\frac{2}{9}-\delta^2-\frac{1}{3}\delta\\
&= \frac{3}{16}\delta^2(3\delta-1)+\frac{1}{4}\leqslant\frac{1}{4}.
\end{aligned}$$

15. When $n=2$, $(a_1+a_2)^2-(a_1-a_2)^2=(b_1+b_2)^2-(b_1-b_2)^2$. Since $a_1-a_2\leqslant b_1-b_2$, $a_1+a_2\leqslant b_1+b_2$.

When $n=3$,

$$\begin{aligned}
&2b_1+2b_2+2b_3-(a_1+a_2+a_3)\\
&=b_1+2b_2+3b_3+(b_1-b_3)-a_2-2a_3-(a_1-a_3)\\
&\geqslant b_1+2b_2+3b_3-a_2-2a_3 &\text{①}\\
&=(b_1-b_3)+2b_2+4b_3-(a_2-a_3)-3a_3\\
&\geqslant 2b_2+4b_3-3a_3. &\text{②}
\end{aligned}$$

(1) If $2b_3\geqslant a_3$, then $2b_2+4b_3\geqslant 6b_3\geqslant 3a_3$, from ② the conclusion holds.

(2) If $b_2\geqslant a_2$, then $b_1+2b_2\geqslant 3b_2\geqslant 3a_2\geqslant a_2+2a_3$, from ① the conclusion holds.

(3) If $2b_3<a_3$, $b_2<a_2$, then $b_1>2a_1$, thus $2(b_1+b_2+b_3)>2b_1>4a_1>a_1+a_2+a_3$, so the conclusion still holds.

For the general case when $n\geqslant 3$, WLOG assume that $b_1b_2\cdots b_n=1$. If $a_1\leqslant n-1$, then $\sum_{i=1}^{n}a_i\leqslant n(n-1)\leqslant(n-1)\sum_{i=1}^{n}b_i$, the conclusion holds.

Next, we assume $a_1>n-1$, then

$$\begin{aligned}
\sum_{1\leqslant i<j\leqslant n}(a_i-a_j) &= \sum_{i=1}^{n}(n-2i+1)a_i=\sum_{i=1}^{n}a_i+\sum_{i=1}^{n}(n-2i)a_i\\
&\geqslant \sum_{i=1}^{n}a_i+(n-2)(a_1-a_{n-1})-na_n\\
&\geqslant \sum_{i=1}^{n}a_i+(a_1-a_{n-1})-na_n\,(n\geqslant 3),
\end{aligned}$$

$$\sum_{1 \leqslant i < j \leqslant n} (b_i - b_j) = \sum_{i=1}^{n} (n - 2i + 1) b_i = \sum_{i=1}^{n} [(n-1)b_i + (2-2i)b_i]$$
$$\leqslant (n-1)\sum_{i=1}^{n} b_i - 2b_2 - 2(n-1)b_n.$$

We now assume $a_1 - a_{n-1} - na_n + 2b_2 + 2(n-1)b_n < 0$, otherwise the conclusion obviously holds. Therefore, $na_n > 2(n-1)b_n + 2b_2 \geqslant 2nb_n$, so $a_n > 2b_n$. On the other hand, since $a_1 a_2 \cdots a_n = 1$, $a_n \leqslant 1$, so $a_1 - (n-1)a_n > (n-1) - (n-1) = 0$. As a result $2b_2 < a_{n-1} + a_n \leqslant 2a_{n-1}$, so $b_2 < a_{n-1}$.

We have $b_1 b_2 \cdots b_n = a_1 a_2 \cdots a_n > 2b_n \cdot b_2 \cdot a_1 a_2 \cdots a_{n-2}$, in other words $b_1 b_3 b_4 \cdots b_{n-1} > 2a_1 a_2 \cdots a_{n-2}$.

Note that $b_3 \leqslant b_2 < a_{n-1} \leqslant a_{n-2}$, $b_4 \leqslant b_3 < a_{n-2} \leqslant a_{n-3}$, $\cdots$, $b_{n-1} \leqslant b_{n-2} < a_3 \leqslant a_2$, so $b_1 > 2a_1$.

We conclude that $(n-1)\sum_{i=1}^{n} b_i > 2(n-1)a_1 > na_1 \geqslant \sum_{i=1}^{n} a_i \, (n \geqslant 3)$.

16. It's easy to see that the original inequality is equivalent to prove

$$\sum_{\text{cyc}} \frac{yz}{x(y+z)^2} \geqslant \frac{9}{4(x+y+z)}.$$

WLOG assume that $x \geqslant y \geqslant z$, and $x + y + z = 1$, then

$$\begin{aligned}
\sum_{\text{cyc}} \frac{4yz}{x(y+z)^2} &= \sum_{\text{cyc}} \frac{(y+z)^2 - (y-z)^2}{x(y+z)^2} \\
&= \sum_{\text{cyc}} \frac{1}{x} - \sum_{\text{cyc}} \frac{(y-z)^2}{x(y+z)^2} \\
&= \sum_{\text{cyc}} \frac{1}{x} \cdot \sum_{\text{cyc}} x - \sum_{\text{cyc}} \frac{(y-z)^2}{x(y+z)^2} \\
&= 9 + \sum_{\text{cyc}} \left(\frac{\sqrt{z}}{\sqrt{y}} - \frac{\sqrt{y}}{\sqrt{z}} \right)^2 - \sum_{\text{cyc}} \frac{(y-z)^2}{x(y+z)^2} \\
&= 9 + \sum_{\text{cyc}} \left(\frac{(y-z)^2}{yz} - \frac{(y-z)^2}{x(y+z)^2} \right) \\
&= 9 + S,
\end{aligned}$$

where $S = \sum_{\text{cyc}} (y-z)^2 \cdot \left(\frac{1}{yz} - \frac{1}{x(y+z)^2}\right)$.

Since $x > \frac{1}{4}$, so

$$\frac{1}{yz} \geqslant \frac{4}{(y+z)^2} > \frac{1}{x(y+z)^2}.$$

Also,

$$\frac{1}{xz} - \frac{1}{y(x+z)^2} \geqslant 0,$$

$$y(x+z)^2 \geqslant xz(x+y+z),$$

$$x^2(y-z) + xz(y-z) + yz^2 \geqslant 0,$$

thus

$$S \geqslant (x-y)^2 \cdot \left(\frac{1}{xz} - \frac{1}{y(x+z)^2} + \frac{1}{xy} - \frac{1}{z(x+y)^2}\right).$$

But

$$\begin{aligned}
&\frac{1}{xz} - \frac{1}{y(x+z)^2} + \frac{1}{xy} - \frac{1}{z(x+y)^2} \\
&= \frac{(x+z)^2 - x}{xy(x+z)^2} + \frac{(x+y)^2 - x}{xz(x+y)^2} \\
&\geqslant \frac{(x+y)^2 - x + (x+z)^2 - x}{xy(x+z)^2} \\
&= \frac{2x^2 + 2xy + 2xz + y^2 + z^2 - 2x(x+y+z)}{xy(x+z)^2} \geqslant 0,
\end{aligned}$$

so the conclusion holds.

17. Let $x = \cot A$, $y = \cot B$, $z = \cot C$. Since $A + B + C = \pi$, $\cot A \cot B + \cot B \cot C + \cot C \cot A = 1$, so $xy + yz + zx = 1$ and the original inequality is equivalent to

$$x^3 + y^3 + z^3 + 6xyz \geqslant (x+y+z)(xy+yz+zx),$$

or

$$x(x-y)(x-z) + y(y-z)(y-x) + z(z-x)(z-y) \geqslant 0.$$

From Schur's Inequality the above inequality holds.

18. Since $\sin(x - y)\sin(x + y) = \frac{1}{2}(\cos 2\beta - \cos 2\alpha) = \sin^2\alpha - \sin^2\beta$,

$$\begin{aligned}&\sin a \sin(a - b)\sin(a - c)\sin(a + b)\sin(a + c)\\ &= \sin a(\sin^2 a - \sin^2 b)(\sin^2 a - \sin^2 c).\end{aligned}$$

Let $x = \sin^2 a$, $y = \sin^2 b$, $z = \sin^2 c$, the original inequality is equivalent to

$$x^{1/2}(x - y)(x - z) + y^{1/2}(y - z)(y - x) + z^{1/2}(z - x)(z - y) \geqslant 0,$$

so Schur's Inequality applies.

19. From the given information, we have $a^2 + b^2 + c^2 + ab + bc + ac \leqslant 2$. Thus,

$$\begin{aligned}\sum_{\text{cyc}} \frac{2ab + 2}{(a + b)^2} &\geqslant \sum_{\text{cyc}} \frac{2ab + a^2 + b^2 + c^2 + ab + bc + ca}{(a + b)^2}\\ &= \sum_{\text{cyc}} \frac{(a + b)^2 + (c + a)(c + b)}{(a + b)^2}\\ &= 3 + \sum_{\text{cyc}} \frac{(c + a)(c + b)}{(a + b)^2} \geqslant 6 \text{ (using AM-GM)}.\end{aligned}$$

Dividing each side by 2 yields the original inequality.

20. Since

$$\begin{aligned}&\frac{a^k}{a + b} + \frac{1}{4}(a + b) + \frac{1}{2} + \frac{1}{2} + \cdots + \frac{1}{2}\left(\frac{1}{2} \text{ is repeated } k - 2 \text{ times}\right)\\ &\geqslant k \cdot \sqrt[k]{\frac{a^k}{2^k}} = \frac{k}{2}a,\end{aligned}$$

we have

$$\frac{a^k}{a + b} \geqslant \frac{k}{2}a - \frac{1}{4}(a + b) - \frac{k - 2}{2}.$$

Similarly,

$$\begin{aligned}\frac{b^k}{b + c} &\geqslant \frac{k}{2}b - \frac{1}{4}(b + c) - \frac{k - 2}{2},\\ \frac{c^k}{c + a} &\geqslant \frac{k}{2}c - \frac{1}{4}(c + a) - \frac{k - 2}{2}.\end{aligned}$$

Adding these three inequalities together, we obtain

$$\frac{a^k}{a+b}+\frac{b^k}{b+c}+\frac{c^k}{c+a}\geqslant\frac{k-1}{2}(a+b+c)-\frac{3}{2}(k-2)$$
$$\geqslant\frac{3}{2}(k-1)-\frac{3}{2}(k-2)=\frac{3}{2}.$$

Exercise 2

1. The original inequality is equivalent to $(a^p-b^p)(a^q-b^q)\geqslant 0$. Since a^p-b^p and a^q-b^q are of the same sign (or they are both zero), this is obviously true.

2. The original inequality is equivalent to $\left(\sum a_i^2b_i\right)\left(\sum a_i\right)-\left(\sum a_i^2\right)\left(\sum a_ib_i\right)\geqslant 0$, or $\frac{1}{2}\sum\sum(a_i-a_j)(b_i-b_j)a_ia_j\geqslant 0$.

3. Assume $a_1\leqslant a_2\leqslant\cdots\leqslant a_n$, then

$$a_j-a_i=(a_j-a_{j-1})+(a_{j-1}-a_{j-2})+\cdots+(a_{i+1}-a_i)\geqslant(j-i)m.$$

Thus $\sum\limits_{1\leqslant i<j\leqslant n}(a_i-a_j)^2\geqslant\frac{1}{12}n^2(n-1)(n+1)m^2$. Also, since

$$\sum_{1\leqslant i<j\leqslant n}(a_i-a_j)^2=n(a_1^2+a_2^2+\cdots+a_n^2)-(a_1+a_2+\cdots+a_n)^2\leqslant n,$$

we have $m\leqslant\sqrt{\frac{12}{(n-1)n(n+1)}}$ and it's easy to find the conditions when the equality holds.

4. Obviously $a_k>a_{k-1}$, and $a_k\geqslant 2$, $k=1, 2, \cdots, n-1$. From the given conditions,

$$\frac{a_{k-1}}{a_k}\leqslant\frac{a_{k-1}}{a_k-1}-\frac{a_k}{a_{k+1}-1}.$$

Sum the above inequality for $k=i+1, i+2, \cdots, n$, we have

$$\frac{a_i}{a_{i+1}}+\frac{a_{i+1}}{a_{i+2}}+\cdots+\frac{a_{n-1}}{a_n}<\frac{a_i}{a_{i+1}-1}.$$

When $i=0$, the above inequality and given conditions imply that $\frac{1}{a_1}\leqslant\frac{99}{100}<\frac{1}{a_1-1}$. Therefore

$$\frac{100}{99} \leqslant a_1 < \frac{100}{99} + 1, \text{ so } a_1 = 2.$$

When $i = 1, 2, 3$, similarly we find $a_2 = 5$, $a_3 = 56$, $a_4 = 78\,400$. Hence $\frac{1}{a_5} \leqslant \frac{1}{a_4}\left(\frac{99}{100} - \frac{1}{2} - \frac{2}{5} - \frac{5}{56} - \frac{56}{25 - 56^2}\right) = 0$. So there is no a_5, and the solution is given by $a_1 = 2$, $a_2 = 5$, $a_3 = 56$, $a_4 = 78\,400$.

5. The original inequality is equivalent to

$$\sum_{k=1}^{n} a_k^{2n} \cdot \sum_{k=1}^{n} \frac{1}{a_k^{2n}} - n^2 \sum_{i<j}\left(\frac{a_i}{a_j} - \frac{a_j}{a_i}\right)^2 > n^2.$$

According to Lagrange's Identity,

$$\sum_{k=1}^{n} a_k^{2n} \cdot \sum_{k=1}^{n} \frac{1}{a_k^{2n}} - n^2 = \sum_{i<j}\left(\frac{a_i^n}{a_j^n} - \frac{a_j^n}{a_i^n}\right)^2.$$

It's easy to check that for $x > 0$, $\left(x^n - \frac{1}{x^n}\right)^2 \geqslant n^2\left(x - \frac{1}{x}\right)^2$, therefore

$$\sum_{k=1}^{n} a_k^{2n} \cdot \sum_{k=1}^{n} \frac{1}{a_k^{2n}} - n^2 \geqslant n^2 \sum_{i<j}\left(\frac{a_i}{a_j} - \frac{a_j}{a_i}\right)^2,$$

and the equality cannot hold.

6. Note that

$$\begin{aligned}
n^2(y - x^2) &= n\sum_{j=1}^{n} a_j^2 - \left(\sum_{j=1}^{n} a_j\right)^2 \\
&= (n-1)\sum_{j=1}^{n} a_j^2 - 2\sum_{1\leqslant i<j\leqslant n} a_i a_j \\
&= (n-1)(a_1^2 + a_n^2) - 2a_1 a_n - 2a_1(a_2 + a_3 + \cdots + a_{n-1}) \\
&\quad - 2a_n(a_2 + a_3 + \cdots + a_{n-1}) + \sum_{2\leqslant i<j\leqslant n-1} (a_i - a_j)^2 \\
&\quad + 2(a_2^2 + a_3^2 + \cdots + a_{n-1}^2) \\
&= \sum_{2\leqslant i<j\leqslant n-1} (a_i - a_j)^2 + 2\sum_{j=2}^{n-1}\left(a_j - \frac{a_1 + a_n}{2}\right)^2 + \frac{n}{2}(a_1 - a_n)^2 \\
&\geqslant \frac{n}{2}(a_n - a_1)^2.
\end{aligned}$$

Thus $a_n - a_1 \leqslant \sqrt{2n(y-x^2)}$.

On the other hand,

$$n^2(y-x^2) + n\sum_{j=2}^{n-1}(a_n - a_j)(a_j - a_1)$$

$$= (n-1)\sum_{j=1}^{n} a_j^2 - 2\sum_{1\leqslant i<j\leqslant n} a_i a_j + na_n\sum_{j=2}^{n-1} a_j + na_1\sum_{j=2}^{n-1} a_j - n(n-2)a_1 a_n - n\sum_{j=2}^{n-1} a_j^2$$

$$= (n-1)(a_1^2 + a_n^2) - \sum_{j=2}^{n-1} a_j^2 + (n-2)a_n\sum_{j=2}^{n-1} a_j + (n-2)a_1\sum_{j=2}^{n-1} a_j - [n(n-2)+2]a_1 a_n - 2\sum_{2\leqslant i<j\leqslant n-1} a_i a_j$$

$$= -\left[\left(\sum_{j=2}^{n-1} a_j\right) - \frac{n-1}{2}(a_1 + a_n)\right]^2 + \frac{n^2}{4}(a_1 - a_n)^2.$$

Hence we have $a_n - a_1 \geqslant 2\sqrt{y-x^2}$. We conclude that

$$2\sqrt{y-x^2} \leqslant a_n - a_1 \leqslant \sqrt{2n(y-x^2)}.$$

7. We have

$$\begin{aligned}
\sum_{k=1}^{n-1} | z'_k - z'_{k+1} | &= \sum_{k=1}^{n-1}\left|\frac{1}{k}\sum_{j=1}^{k} z_j - \frac{1}{k+1}\sum_{j=1}^{k+1} z_j\right| \\
&= \sum_{k=1}^{n-1}\frac{1}{k(k+1)}\left|\sum_{j=1}^{k} j(z_j - z_{j+1})\right| \\
&\leqslant \sum_{k=1}^{n-1}\left(\frac{1}{k(k+1)}\cdot\sum_{j=1}^{k} j\,|z_j - z_{j+1}|\right) \\
&= \sum_{j=1}^{n-1}\sum_{k=j}^{n-1}\left(\frac{1}{k(k+1)} j\,|z_j - z_{j+1}|\right) \\
&= \sum_{j=1}^{n-1}\left(1 - \frac{j}{n}\right)|z_j - z_{j+1}| \\
&\leqslant \sum_{j=1}^{n-1}\left(1 - \frac{1}{n}\right)|z_j - z_{j+1}| \\
&= \left(1 - \frac{1}{n}\right)\sum_{j=1}^{n-1}|z_j - z_{j+1}|.
\end{aligned}$$

It's easy to see that when $z_1 \neq z_2$, $z_2 = z_3 = \cdots = z_n$ the equality holds.

8. Let $y_i = \sum_{j=i+1}^{n} x_j$, $y = \sum_{j=2}^{n} (j-1)x_j$, $z_i = \frac{n(n-1)}{2} y_i - (n-i)y (1 \leqslant i \leqslant n-1)$. Thus,

$$\begin{aligned} &\frac{n(n-1)}{2} \cdot \sum_{1 \leqslant i < j \leqslant n} x_i x_j - \left[\sum_{i=1}^{n-1} (n-i)x_i\right] \cdot \left[\sum_{j=2}^{n} (j-1)x_j\right] \\ &= \frac{n(n-1)}{2} \cdot \sum_{i=1}^{n-1} x_i \sum_{j=i+1}^{n} x_j - \sum_{i=1}^{n-1} (n-i)x_i y \\ &= \sum_{i=1}^{n-1} x_i \cdot \left[\frac{n(n-1)}{2} y_i - (n-i)y\right] = \sum_{i=1}^{n-1} x_i z_i. \end{aligned}$$

It suffices to show that $\sum_{i=1}^{n-1} x_i z_i > 0$.

Note that $\sum_{i=1}^{n-1} y_i = y$, $\sum_{i=1}^{n-1} z_i = 0$, $y = \sum_{j=2}^{n} (j-1)x_j < \sum_{j=2}^{n} (j-1)x_n = \frac{n(n-1)}{2} x_n$, $z_{n-1} = \frac{n(n-1)}{2} y_{n-1} - y = \frac{n(n-1)}{2} x_n - y > 0$.

Also, since $\frac{z_{i+1}}{\frac{n(n-1)}{2}(n-i-1)} - \frac{z_i}{\frac{n(n-1)}{2}(n-i)} = \frac{y_{i+1}}{n-i-1} - \frac{y_i}{n-i} = \frac{x_{i+2} + x_{i+3} + \cdots + x_n}{n-i-1} - \frac{x_{i+1} + x_{i+2} + \cdots + x_n}{n-i} > 0$, we have $\frac{z_1}{n-1} < \frac{z_2}{n-2} < \cdots < \frac{z_{n-2}}{2} < z_{n-1}$.

Hence there must exist a positive integer k, when $1 \leqslant i \leqslant k$, $z_i \leqslant 0$, and when $k+1 \leqslant i \leqslant n$, $z_i > 0 (k \leqslant n-2)$, so $(x_i - x_k)z_i \geqslant 0$.

As a result, $\sum_{i=1}^{n-1} x_i z_i \geqslant x_k \sum_{i=1}^{n-1} z_i = 0$, and the equality cannot hold.

9. We have

$$\begin{aligned} &\left[\sum_{k=1}^{n} (\sqrt{k} - \sqrt{k-1}) \sqrt{a_k}\right]^2 \\ &= \sum_{i=1}^{n} (\sqrt{i} - \sqrt{i-1})^2 a_i + 2\sum_{i<j} (\sqrt{i} - \sqrt{i-1})(\sqrt{j} - \sqrt{j-1}) \sqrt{a_i a_j} \end{aligned}$$

$$\geqslant \sum_{i=1}^{n}(\sqrt{i}-\sqrt{i-1})^{2}a_{i}+2\sum_{i<j}(\sqrt{i}-\sqrt{i-1})(\sqrt{j}-\sqrt{j-1})a_{j}$$

$$=\sum_{i=1}^{n}(\sqrt{i}-\sqrt{i-1})^{2}a_{i}+2\sum_{j=1}^{n}\sqrt{j-1}(\sqrt{j}-\sqrt{j-1})a_{j}$$

$$=\sum_{i=1}^{n}a_{i}.$$

Therefore the conclusion holds. It's easy to see that the equality holds if there exists m, such that $a_1 = a_2 = \cdots = a_m$ and $a_k = 0$ when $k > m$.

Note: In the following we give two more proves.

Proof 2: using Abel's summation formula, we transform the inequality into

$$\sqrt{\sum_{k=1}^{n}a_k} \leqslant \sum_{k=1}^{n}\sqrt{a_k}(\sqrt{k}-\sqrt{k-1}) = \sum_{k=1}^{n}\sqrt{a_k c_k},$$

where $c_k = \sqrt{k}-\sqrt{k-1}$.

Since $a_1 \geqslant a_2 \geqslant \cdots \geqslant a_n$, $c_1 \geqslant c_2 \geqslant \cdots \geqslant c_n$, we are motivated to try Chebyshev's Inequality (see problem 10). That is,

$$\sum_{k=1}^{n}\sqrt{a_k}\cdot c_k \geqslant \frac{1}{n}\cdot\sum_{k=1}^{n}\sqrt{a_k}\cdot\sum_{k=1}^{n}c_k = \frac{1}{\sqrt{n}}\sum_{k=1}^{n}\sqrt{a_k},$$

which is unfortunately not helpful at all. Now if we change the conclusion to the following:

$$\sqrt{\sum_{k=1}^{n}a_k} \leqslant \sum_{k=1}^{n-1}\sqrt{k}(\sqrt{a_k}-\sqrt{a_{k+1}})+\sqrt{na_n}, \tag{①}$$

we can use induction on n to prove the result. In fact ① is obviously true for $n = 1$.

Assume ① holds for some $n \geqslant 1$ and every non-negative and non-increasing series of length n. Consider a series with length $n+1$, which satisfies $a_1 \geqslant a_2 \geqslant \cdots \geqslant a_{n+1} \geqslant 0$. From the induction assumptions, it suffices to show that $\sqrt{\sum_{k=1}^{n+1}a_k} - \sqrt{\sum_{k=1}^{n}a_k} \leqslant -\sqrt{na_{n+1}} + \sqrt{(n+1)a_{n+1}}$.

WLOG assume that $a_{n+1} > 0$, $S = \sum_{k=1}^{n} a_k$, $m = \dfrac{S}{a_{n+1}}$, the above is then equivalent to $\sqrt{m+1} - \sqrt{m} \leqslant \sqrt{n+1} - \sqrt{n}$, which is obviously true.

Proof 3: Let $x_i = \sqrt{a_i} - \sqrt{a_{i+1}}$, $i = 1, 2, \cdots, n$. We have $a_i = (x_i + x_{i+1} + \cdots + x_n)^2$, $1 \leqslant i \leqslant n$. Hence $\sum_{k=1}^{n} a_k = \sum_{k=1}^{n} kx_k^2 + 2\sum_{1\leqslant k<l\leqslant n} kx_k x_l \leqslant \sum_{k=1}^{n} kx_k^2 + 2\sum_{k<l} \sqrt{kl}\, x_k x_l = \left(\sum_{k=1}^{n} \sqrt{k}\, x_k\right)^2$, and the conclusion follows.

10. Denote $b_{n+t} = b_n$, $t = 0, 1, 2, \cdots$. Then $\sum_{i=1}^{n} b_{i+t} = \sum_{i=1}^{n} b_i$, $\sum_{i=1}^{k} b_{i+t} \geqslant \sum_{i=1}^{k} b_i$, $a_k - a_{k+1} \leqslant 0 (1 \leqslant k \leqslant n-1)$.

Using summation by parts, we have

$$\begin{aligned}\left(\sum_{k=1}^{n} a_k\right)\left(\sum_{k=1}^{n} b_k\right) &= \sum_{t=0}^{n-1}\left(\sum_{i=1}^{n} a_i b_{i+t}\right)\\ &= \sum_{t=0}^{n-1}\left[a_n \sum_{i=1}^{n} b_{i+t} + \sum_{k=1}^{n-1}\left(\sum_{i=1}^{k} b_{i+t}\right)(a_k - a_{k+1})\right]\\ &\leqslant \sum_{t=0}^{n-1}\left[a_n \sum_{i=1}^{n} b_i + \sum_{k=1}^{n-1}\left(\sum_{i=1}^{k} b_i\right)(a_k - a_{k+1})\right]\\ &= n \cdot \sum_{k=1}^{n} a_k b_k.\end{aligned}$$

Similarly we can show that $\left(\sum_{k=1}^{n} a_k\right)\left(\sum_{k=1}^{n} b_k\right) \geqslant n \sum_{k=1}^{n} a_k b_{n-k+1}$.

11. From symmetry, WLOG we assume that $a_1 \geqslant a_2 \geqslant \cdots \geqslant a_n$, then

$$\sum_{j=1}^{k} a_j^q \geqslant \sum_{j=1}^{k} a_{i_j}^q \ (1 \leqslant k \leqslant n-1), \ \sum_{j=1}^{n} a_j^q = \sum_{j=1}^{n} a_{i_j}^q,$$

and $(S^p - a_k^p)^{-1} \geqslant (S^p - a_{k+1}^p)^{-1} (1 \leqslant k \leqslant n-1)$.

According to Abel's summation by parts formula,

$$\begin{aligned}\sum_{k=1}^{n} \frac{a_k^q - a_{i_k}^q}{S^p - a_k^p} &= (S^p - a_n^p)^{-1} \cdot \sum_{j=1}^{n} (a_j^q - a_{i_j}^q)\\ &\quad + \sum_{k=1}^{n-1}\left[\sum_{j=1}^{k} (a_j^q - a_{i_j}^q)\right] \cdot \left[(S^p - a_k^p)^{-1} - (S^p - a_{k+1}^p)^{-1}\right]\\ &\geqslant 0.\end{aligned}$$

So the original inequality holds.

12. Denote $b_k = a_1 + a_2 + \cdots + a_k - \sqrt{k} \geqslant 0$, $1 \leqslant k \leqslant n$, thus $a_k = (b_k - b_{k-1}) + (\sqrt{k} - \sqrt{k-1})$. As a result,

$$\begin{aligned}
\sum_{j=1}^{n} a_j^2 &= [b_1^2 + (b_2 - b_1)^2 + \cdots + (b_n - b_{n-1})^2] + [1^2 + (\sqrt{2} - \sqrt{1})^2 + \cdots \\
&\quad + (\sqrt{n} - \sqrt{n-1})^2] + [2 \cdot \sqrt{1} \cdot b_1 + 2 \cdot (\sqrt{2} - \sqrt{1}) \cdot (b_2 - b_1) \\
&\quad + \cdots + 2 \cdot (\sqrt{n} - \sqrt{n-1}) \cdot (b_n - b_{n-1})] \\
&\geqslant [1^2 + (\sqrt{2} - \sqrt{1})^2 + \cdots + (\sqrt{n} - \sqrt{n-1})^2] + [2 \cdot \sqrt{1} \cdot b_1 + 2(\sqrt{2} - \\
&\quad \sqrt{1})(b_2 - b_1) + \cdots + 2(\sqrt{n} - \sqrt{n-1})(b_n - b_{n-1})] \\
&\geqslant 1^2 + (\sqrt{2} - \sqrt{1})^2 + \cdots + (\sqrt{n} - \sqrt{n-1})^2 \\
&> \frac{1}{4} \cdot \left(1 + \frac{1}{2} + \cdots + \frac{1}{n}\right).
\end{aligned}$$

13. Let $A_n = \sum_{k=1}^{n} \frac{[kx]}{k}$, we use induction to show that $A_n \leqslant [nx]$.

When $n = 1$, this is obviously true. Now assume for $1 \leqslant k \leqslant n-1$, $A_k \leqslant [kx]$.

From the summation by parts formula,

$$\begin{aligned}
nA_n &= \sum_{k=1}^{n} k \cdot \frac{[kx]}{k} - \sum_{k=1}^{n-1} A_k (k - (k+1)) \\
&= \sum_{k=1}^{n} [kx] + \sum_{k=1}^{n-1} A_k \leqslant \sum_{k=1}^{n} [kx] + \sum_{k=1}^{n-1} [kx] \\
&= [nx] + \sum_{k=1}^{n-1} ([n-k]x + [kx]) \\
&\leqslant [nx] + \sum_{k=1}^{n-1} [(n-k)x + kx] (\text{since} [x] + [y] \leqslant [x+y]) \\
&= n[nx].
\end{aligned}$$

Thus $A_n \leqslant [nx]$.

14. If we assume $\frac{1}{c} \sum_{k=1}^{n} A_k^2 \leqslant \sum_{k=1}^{n} A_k \cdot a_k$, we have $\sum_{k=1}^{n} A_k \cdot a_k \leqslant c \cdot \sum_{k=1}^{n} a_k^2$, and Abel's formulas can be used as a result:

$$\begin{aligned}\sum_{k=1}^{n} A_k a_k &= \sum_{k=1}^{n} A_k[kA_k - (k-1)A_{k-1}] \\ &= \sum_{k=1}^{n} kA_k^2 - \sum_{k=1}^{n}(k-1)A_kA_{k-1} \\ &\geqslant \sum_{k=1}^{n} kA_k^2 - \frac{1}{2}\Big[\sum_{k=1}^{n}(k-1)A_k^2 + \sum_{k=1}^{n}(k-1)A_{k-1}^2\Big] \\ &= \frac{1}{2}\cdot\sum_{k=1}^{n} A_k^2 + \frac{1}{2}nA_n^2 \geqslant \frac{1}{2}\sum_{k=1}^{n} A_k^2.\end{aligned}$$

Therefore $\sum_{k=1}^{n} A_k^2 \leqslant 4\sum_{k=1}^{n} a_k^2$.

Comment: We can use the same idea to prove Hölder type inequality as the following: assume $p > 1$, then

$$\sum_{k=1}^{n} A_k^p \leqslant \left(\frac{p}{p-1}\right)^p \cdot \sum_{k=1}^{n} a_k^p.$$

Similarly, we have

$$\frac{1}{c}\sum_{k=1}^{n} A_k^p \leqslant \sum_{k=1}^{n} A_k^{p-1}\cdot a_k \leqslant c^{p-1}\sum_{k=1}^{n} a_k^p,$$

and we can show that setting the undetermined coefficient c to $\frac{p}{p-1}$ yields the desired inequality. This problem can be regarded as an elementary form of Hardy-Landau type inequalities.

Exercise 3

1. Let $b+3c=x$, $8c+4a=y$, $3a+2b=z$, then the original sum is equivalent to

$$\frac{1}{8}\left(\frac{y}{x}+\frac{4x}{y}\right)+\frac{1}{6}\left(\frac{z}{x}+\frac{9x}{z}\right)+\frac{1}{16}\left(\frac{4z}{y}+\frac{9y}{z}\right)\geqslant\frac{47}{48}.$$

2. WLOG assume that $m, n \in \mathbf{R}^+$, $a, b \in \mathbf{R}^+$. Let $a = m\cos\alpha$, $b = n\sin\alpha$, where $\alpha \in \left(0, \frac{\pi}{2}\right]$, thus $(m+n)^2 = \left(\frac{a}{\cos\alpha}+\frac{b}{\sin\alpha}\right)^2 =$

$\frac{a^2}{\cos^2\alpha}+\frac{b^2}{\sin^2\alpha}+\frac{2ab}{\sin\alpha\cos\alpha}=a^2(1+\tan^2\alpha)+b^2\left(1+\frac{1}{\tan^2\alpha}\right)+\frac{2ab(1+\tan^2\alpha)}{\tan\alpha}$.

Denote $\tan\alpha=t$, so $t\in\mathbf{R}^+$, thus

$$\begin{aligned}(m+n)^2&=(a^2+b^2)+a^2t^2+\frac{b^2}{t^2}+\frac{2ab}{t}+2abt\\&=a\left(at^2+\frac{b}{t}+\frac{b}{t}\right)+b\left(\frac{b}{t^2}+at+at\right)+(a^2+b^2)\\&\geqslant a\cdot 3\sqrt[3]{ab^2}+b\cdot 3\sqrt[3]{a^2b}+(a^2+b^2).\end{aligned}$$

Therefore $(m+n)_{\min}=(a^{2/3}+b^{2/3})^{3/2}$, and the equality holds when $t=\sqrt[3]{\frac{b}{a}}$.

Comment: We can also use Cauchy's Inequality.

Since $(a^{1/3}\cos\alpha+b^{1/3}\sin\alpha)^2\leqslant(a^{2/3}+b^{2/3})(\sin^2\alpha+\cos^2\alpha)$, we have $(a^{2/3}+b^{2/3})^{1/2}\left(\frac{a}{\cos\alpha}+\frac{b}{\sin\alpha}\right)\geqslant(a^{1/3}\cos\alpha+b^{1/3}\sin\alpha)\left(\frac{a}{\cos\alpha}+\frac{b}{\sin\alpha}\right)\geqslant(a^{2/3}+b^{2/3})^2$. Thus $m+n\geqslant(a^{2/3}+b^{2/3})^{3/2}$, and the equality holds when $\alpha=\arctan\sqrt[3]{\frac{a}{b}}$.

3. Assume $b+c-a=2x$, $c+a-b=2y$, $a+b-c=2z$, then $x, y, z\in\mathbf{R}^+$, and $a=y+z$, $b=x+z$, $c=x+y$. So the original inequality is equivalent to $\frac{x-z}{2x}+\frac{y-x}{2y}+\frac{z-y}{2z}\leqslant 0$, or $\frac{z}{x}+\frac{x}{y}+\frac{y}{z}\geqslant 3$, which is apparently true.

4. Assume $x+y=2a$, $y+z=2b$, $z+x=2c$, then a, b, c can form a triangle. Assume its area is S and its circumscribed circle has radius R, then obviously the original inequality is equivalent to $a+b+c\leqslant 3\sqrt{3}R$.

Since $2R(\sin A+\sin B+\sin C)=a+b+c\leqslant 2R\cdot 3\cdot\sin\frac{A+B+C}{3}=3\sqrt{3}R$, the original inequality holds.

5. Assume $x + y = a$, $xy = b$, then $a^2 \geqslant 4b > 0$. Thus

$$\text{LHS} = \left(\frac{2a^2 + b}{9}\right)^2 = \left(\frac{\frac{1}{4}a^2 + \frac{1}{4}a^2 + \cdots + \frac{1}{4}a^2 + b}{9}\right)^2$$

$$\geqslant \sqrt[9]{\frac{1}{4^{16}} \cdot a^{32} \cdot b^2} \geqslant \sqrt[9]{\frac{1}{2^{27}} \cdot a^{27} \cdot b^{9/2}}$$

$$= \frac{1}{8}a^3 \cdot b^{\frac{1}{2}} = \text{RHS}.$$

6. (1) Assume $ab = x > 0$, $a + b = y$, then $y^2 \geqslant 4x$. Thus the RHS of the original inequality $= \frac{y^2 + 8x}{12} = \frac{y^2}{12} + \frac{x}{3} + \frac{x}{3} \geqslant 3 \cdot \sqrt[3]{\frac{x^2y^2}{12 \cdot 3^2}} = \sqrt[3]{\frac{x^2y^2}{4}} =$ LHS, so the original inequality holds. The equality is achieved when $a = b$.

(2) When $x \geqslant 0$, since $\frac{y^2 + 8x}{12} \leqslant \frac{y^2 - x}{3}$, the conclusion follows and the equality holds if and only if $a = b$.

When $x < 0$, $-x > 0$, so the RHS $= \frac{y^2 - x}{3} = \frac{y^2}{3} - \frac{x}{6} - \frac{x}{6} \geqslant 3 \cdot \sqrt[3]{\frac{y^2}{3} \cdot \left(-\frac{x}{6}\right)^2} = \sqrt[3]{\frac{x^2y^2}{4}}$, therefore the inequality still holds. In this case the equality is achieved when $b = -2a$ or $a = -2b$.

7. Let $x = a + b + c \geqslant 3\sqrt[3]{abc} = 3$, $y = \frac{1}{a} + \frac{1}{b} + \frac{1}{c} = ab + bc + ca \geqslant 3\sqrt[3]{a^2b^2c^2} = 3$. The original inequality is equivalent

$$\frac{(1+b+c)(1+c+a) + (1+c+a)(1+a+b) + (1+a+b)(1+b+c)}{(1+a+b)(1+b+c)(1+c+a)}$$

$$\leqslant \frac{(2+b)(2+c) + (2+c)(2+a) + (2+a)(2+b)}{(2+a)(2+b)(2+c)},$$

or

$$\frac{3 + 4x + y + x^2}{2x + x^2 + y + xy} \leqslant \frac{12 + 4x + y}{9 + 4x + y}.$$

The above is equivalent $\left(\frac{5x^2y}{3}-5x^2\right)+\left(\frac{xy^2}{3}-y^2\right)+\left(\frac{4x^2y}{3}-12x\right)+(4xy-12x)+\left(\frac{xy^2}{3}-3y\right)+\left(\frac{xy^2}{3}-9\right)+(2xy-18)\geqslant 0.$

Note that $xy\geqslant 9$, thus the above inequality holds and the result follows.

8. Let $a=\frac{y}{x}$, $b=\frac{z}{y}$, $c=\frac{u}{z}$, $d=\frac{v}{u}$, $e=\frac{x}{v}$, $x, y, z, u, v\in \mathbf{R}^+$, then the original inequality is equivalent to

$$\frac{u+y}{x+z+v}+\frac{z+v}{x+y+u}+\frac{x+u}{y+z+v}+\frac{y+v}{x+z+u}+\frac{x+z}{y+u+v}\geqslant\frac{10}{3}.$$

Add 5 to each side and multiply by 3, the above inequality is equivalent to

$$3(x+y+z+u+v)\left(\frac{1}{x+z+v}+\frac{1}{x+y+u}+\frac{1}{y+z+v}+\frac{1}{x+z+u}+\frac{1}{y+u+v}\right)\geqslant 25,$$

which is obviously true by Cauchy's Inequality.

9. WLOG assume that $a\geqslant b\geqslant c$, and let $\frac{a}{c}=x$, $\frac{b}{c}=y$, hence $x\geqslant y\geqslant 1$.

The original inequality is equivalent to

$$x^2+y^2+1\geqslant\frac{x^2+y^2}{x+y}+\frac{y(1+x^2)}{1+x}+\frac{x(1+y^2)}{1+y}.$$

Removing the denominators, the above is rearranged into

$$(x^4y+xy^4)+(x^4+y^4+x+y)\geqslant(x^3y^2+x^2y^3)+(x^3+y^3+x^2+y^2),$$

or equivalently,

$$xy(x+y)(x-y)^2+x(x+1)(x-1)^2+y(y+1)(y-1)^2\geqslant 0,$$

which clearly holds.

Comment: we also give a direct proof in the following.

Assume $a \geqslant b \geqslant c$,

$$a^2 - \frac{a(b^2+c^2)}{b+c} = \frac{a^2b+a^2c-ab^2-ac^2}{b+c} = \frac{ab(a-b)}{b+c} + \frac{ac(a-c)}{b+c}.$$

Since $\frac{1}{b+c} \geqslant \frac{1}{a+c} \geqslant \frac{1}{a+b}$,

$$\text{LHS}-\text{RHS} = \left[\frac{ab(a-b)}{b+c} - \frac{ab(a-b)}{c+a}\right] + \left[\frac{ac(a-c)}{b+c} - \frac{ac(a-c)}{a+b}\right]$$
$$+ \left[\frac{bc(b-c)}{c+a} - \frac{bc(b-c)}{a+b}\right] \geqslant 0,$$

so the original inequality holds.

10. From the given conditions, $x + (-y) + z = x \cdot (-y) \cdot z$. Assume $x = \tan\alpha$, $y = \tan\beta$, $z = \tan\gamma$, $(\alpha+\beta+\gamma = k\pi)$. Therefore, $p = 2\cos^2\alpha - 2\cos^2\beta + 3\cos^2\gamma = 2\cos^2\alpha - 2\cos^2\beta + 3\cos^2(\alpha+\beta) = -2\sin(\alpha+\beta)\sin(\alpha-\beta) + 3\cos^2\gamma \leqslant 2\sin\gamma + 3 - 3\sin^2\gamma = -3\left(\sin\gamma - \frac{1}{3}\right)^2 + \frac{10}{3} \leqslant \frac{10}{3}$, thus $p_{\max} = \frac{10}{3}$.

11. Let $a = \cos\alpha$, $b = \cos\beta$, α, $\beta \in \left(0, \frac{\pi}{2}\right)$, thus

$$ab + \sqrt{(1-a^2)(1-b^2)} = \cos(\alpha-\beta).$$

Take the square on each side, we have

$$\sqrt{(1-a^2)(1-b^2)} = \frac{1}{2ab} \cdot [\cos^2(\alpha-\beta) - 1] + \frac{a}{2b} + \frac{b}{2a} - ab.$$

When $0 < |\alpha - \beta| < \frac{\pi}{6}$, $\cos(\alpha-\beta) > \frac{\sqrt{3}}{2}$, hence

$$\frac{1}{2ab} \cdot [\cos^2(\alpha-\beta) - 1] > -\frac{1}{8ab},$$

so the original inequality holds.

12. Assume $a = \max\{a, b, c\}$, then $AP = \min\{AP, BQ, CR\}$, from the given conditions we have

$$\frac{4}{-a+b+c}+\frac{10}{-b+a+c}+\frac{10}{-c+a+b}=\frac{6}{r}.$$

Let $-a+b+c=2x$, $-b+a+c=2y$, $-c+a+b=2z$, x, y, $z>0$. The above is equivalent to

$$\frac{2}{x}+\frac{5}{y}+\frac{5}{z}=6\cdot\sqrt{\frac{1}{xy}+\frac{1}{yz}+\frac{1}{zx}},$$

so
$$2\left(\frac{1}{x}-\frac{4}{y}\right)^2+2\left(\frac{1}{x}-\frac{4}{z}\right)^2=7\left(\frac{1}{y}-\frac{1}{z}\right)^2.$$

Let $p=\frac{1}{x}-\frac{4}{y}$, $q=\frac{1}{x}-\frac{4}{z}$, then $25p^2+14pq+25q^2=0$. It's easy to see that $p=q=0$, hence $y=z=4x$, so the conclusion holds.

13. Let $a=\frac{x}{y}$, $b=\frac{y}{z}$, $c=\frac{z}{x}$, and x, y, $z\in\mathbf{R}^+$, the original inequality is thus equivalent to

$$\left(\frac{x}{y}-1+\frac{z}{y}\right)\left(\frac{y}{z}-1+\frac{x}{z}\right)\left(\frac{z}{x}-1+\frac{y}{x}\right)\leqslant 1,$$

or $(x-y+z)(y-z+x)(z-x+y)\leqslant xyz$. Any two among $x-y+z$, $y-z+x$, $z-x+y$ have a positive sum, thus at most one of these is not positive.

(1) If exactly one of $x-y+z$, $y-z+x$, $z-x+y$ is $\leqslant 0$, the conclusion obviously follows.

(2) If each one is >0, since $(x-y+z)(y-z+x)\leqslant x^2$, $(x-y+z)(z-x+y)\leqslant z^2$, $(y-z+x)(z-x+y)\leqslant y^2$, multiply these together yields the result.

14. Denote $x=\frac{a}{\sqrt{a^2+8bc}}$, $y=\frac{b}{\sqrt{b^2+8ac}}$, $z=\frac{c}{\sqrt{c^2+8ab}}$, x, y, $z\in\mathbf{R}^+$, then

$$\left(\frac{1}{x^2}-1\right)\left(\frac{1}{y^2}-1\right)\left(\frac{1}{z^2}-1\right)=512.$$

Assume the contrary that $x+y+z<1$, so $0<x$, y, $z<1$, as a result

$$\left(\frac{1}{x^2}-1\right)\left(\frac{1}{y^2}-1\right)\left(\frac{1}{z^2}-1\right)=\frac{(1-x^2)(1-y^2)(1-z^2)}{x^2y^2z^2}$$

$$>\frac{[(x+y+z)^2-x^2]\cdot[(x+y+z)^2-y^2]\cdot[(x+y+z)^2-z^2]}{x^2y^2z^2}$$

$$=\frac{(y+z+x+x)(y+z)(x+y+y+z)(x+z)(x+y+z+z)(x+y)}{x^2y^2z^2}$$

$$\geqslant\frac{4\sqrt[4]{x^2yz}\cdot 2\sqrt{yz}\cdot 4\sqrt[4]{xy^2z}\cdot 2\sqrt{xz}\cdot 4\sqrt[4]{xyz^2}\cdot 2\sqrt{xy}}{x^2y^2z^2}=512,$$

which leads to a contradiction. Hence $x+y+z\geqslant 1$.

Comment: we can also show that

$$\frac{a}{\sqrt{a^2+8bc}}\geqslant\frac{a^{4/3}}{a^{4/3}+b^{4/3}+c^{4/3}},$$

$$\frac{b}{\sqrt{b^2+8ca}}\geqslant\frac{b^{4/3}}{a^{4/3}+b^{4/3}+c^{4/3}},$$

$$\frac{c}{\sqrt{c^2+8ab}}\geqslant\frac{c^{4/3}}{a^{4/3}+b^{4/3}+c^{4/3}}.$$

Adding these together yields the desired inequality.

Exercise 4

1. Assume the contrary. Since $abc>0$, WLOG assume $a<0$, $b<0$, $c>0$. As $a+b+c>0$, we have $c>|a+b|$, but then $ab>c\cdot|a+b|$, so $ab>|a+b|^2$, which leads to a contradiction.

2. Assume to the contrary that there exists real numbers x, y, z, t satisfying the set of inequalities. Take the square on each side, we have $(x+y-z+t)(x-y+z-t)>0$, $(y+x-z+t)(y-x+z-t)>0$, $(z+x-y+t)(z-x+y-t)>0$, $(t+x-y+z)(t-x+y-z)>0$. Hence $-(x+y-z+t)^2(x-y+z-t)^2(y-x+z-t)^2(z+x-y+t)^2>0$, which leads to a contradiction.

3. If $bd>q^2$, then $4abcd=4(ac)(bd)>4p^2q^2=(ab+cd)^2=a^2b^2+2abcd+c^2d^2$, so $(ab-cd)^2<0$, which leads to a contradiction.

4. If $a^2+b^2+c^2<abc$, then $a<bc$, $b<ca$, $c<ab$. Thus $a+$

$b + c < ab + bc + ca \leqslant a^2 + b^2 + c^2 < abc$, which is a contradiction.

5. Assume to the contrary that for any $x, y \in [0, 1]$, $| xy - f(x) - g(y) | < \frac{1}{4}$. Set $(x, y) = (0, 0), (0, 1), (1, 0), (1, 1)$ alternatively, we have $| f(0) + g(0) | < \frac{1}{4}$, $| f(0) + g(1) | < \frac{1}{4}$, $| f(1) + g(0) | < \frac{1}{4}$, $| 1 - f(1) - g(1) | < \frac{1}{4}$.

As a result,

$$\begin{aligned} 1 &= | 1 - f(1) - g(1) + f(0) + g(1) + f(1) + g(0) - f(0) - g(0) | \\ &\leqslant | 1 - f(1) - g(1) | + | f(0) + g(1) | + | f(1) + g(0) | \\ &\quad + | f(0) + g(0) | < 1, \end{aligned}$$

which leads to a contradiction.

6. Assume the contrary, then there exists real numbers a, b such that for any $x, y \in [0, 1]$, $| xy - ax - by | < \frac{1}{3}$. Set $(x, y) = (1, 0), (0, 1), (1, 1)$ alternatively, we have $| a | < \frac{1}{3}$, $| b | < \frac{1}{3}$, $| 1 - a - b | < \frac{1}{3}$, then $1 = | a + b + 1 - a - b | \leqslant | a | + | b | + | 1 - a - b | < 1$, which leads to a contradiction.

7. WLOG assume that $a_1 > a_2 > \cdots > a_m$. We next show that for any integer i satisfying $1 \leqslant i \leqslant m$, we have

$$a_i + a_{m+1-i} \geqslant n + 1. \quad ①$$

Note if ① holds, then $2(a_1 + a_2 + \cdots + a_m) = (a_1 + a_m) + (a_2 + a_{m-1}) + \cdots + (a_m + a_1) \geqslant m(n + 1)$, and the conclusion follows.

If there exists an integer i, $1 \leqslant i \leqslant m$, such that $a_i + a_{m+1-i} \leqslant n$, then $a_i < a_i + a_m < a_i + a_{m-1} < \cdots < a_i + a_{m+1-i} \leqslant n$. $a_i + a_m$, $a_i + a_{m-1}$, $\cdots$, $a_i + a_{m+1-i}$ are i integers, and each of them is $\leqslant n$. From the given conditions, each of these i integers can be represented by distinct $a_k (1 \leqslant k \leqslant m)$ larger than a_i, so they must be one of a_1, a_2, $\cdots$, a_{i-1}, which leads to a contradiction.

8. Assume the contrary. If for any $k \in \{1, 2, \cdots, n\}$, we have $S_k = \left|\sum_{i=1}^{k} a_i - \sum_{i=k+1}^{n} a_i\right| > \max_{1 \leqslant i \leqslant n} |a_i|$. Denote $A = \max_{1 \leqslant i \leqslant n} |a_i|$. In addition, denote $S_0 = -S_n$, then one of S_0 and S_n is $> A$, the other is $< -A$. WLOG assume $S_0 < -A$, $S_n > A$ (otherwise we can replace a_i by $-a_i$). Hence, there exists j, $0 \leqslant j \leqslant n-1$, such that $S_j < -A$, $S_{j+1} > A$. So $|S_{j+1} - S_j| > 2A$, or $2|a_{j+1}| > 2A$, then $|a_{j+1}| > A$, which leads to a contradiction.

9. Assume $a + ib$ is a square root of $j_1^2 + j_2^2 + \cdots + j_n^2$, then $r = |a|$, and $(a + ib)^2 = \sum_{k=1}^{n} j_k^2 = \sum_{k=1}^{n} (x_k + iy_k)^2$.

Therefore $a^2 - b^2 = \sum_{k=1}^{n} x_k^2 - \sum_{k=1}^{n} y_k^2$, $ab = \sum_{k=1}^{n} x_k y_k$.

Assume to the contrary that $r > \sum_{k=1}^{n} |x_k|$, then $a^2 = r^2$, $\left(\sum_{k=1}^{n} |x_k|\right)^2 \geqslant \sum_{k=1}^{n} x_k^2$. Thus $b^2 > \sum_{k=1}^{n} y_k^2$, and $a^2 b^2 > \sum_{k=1}^{n} x_k^2 \cdot \sum_{k=1}^{n} y_k^2 \geqslant \left(\sum_{k=1}^{n} x_k y_k\right)^2 = a^2 b^2$, which leads to a contradiction.

10. Assume the contrary. If we cannot find such k numbers, then for $a_1, a_2, \cdots, a_k$, $a_k \leqslant \frac{1}{2} a_1$; for $a_k, a_{k+1}, \cdots, a_{2k-1}$, $a_{2k-1} \leqslant \frac{1}{2} a_k \leqslant \frac{1}{2^2} a_1$; $\cdots$; for $a_{(n-1)(k-1)+1}, a_{(n-1)(k-1)+2}, \cdots, a_{n(k-1)+1}$, $a_{n(k-1)+1} \leqslant \frac{1}{2^n} a_1$. Therefore,

$$S_1 = a_1 + a_k + a_{2k-1} + \cdots \leqslant a_1 + \frac{1}{2} a_1 + \frac{1}{2^2} a_1 + \cdots = 2a_1;$$

$$S_2 = a_2 + a_{k+1} + a_{2k} + \cdots \leqslant 2a_2 \leqslant 2a_1;$$

$$\cdots\cdots$$

$$S_{k-1} = a_{k-1} + a_{2k-2} + \cdots \leqslant 2a_{k-1} \leqslant 2a_1.$$

Hence $S = S_1 + S_2 + \cdots + S_{k-1} \leqslant 2(k-1)a_1 = \frac{k-1}{k} < 1$, which leads to a contradiction.

11. Assume to the contrary that $y(y-1) > x^2$, since $y \geqslant 0$, $y >$

1. In addition, $y > \frac{1}{2} + \sqrt{\frac{1}{4} + x^2}$. From the assumptions that $y(y + 1) \leqslant (x+1)^2$ and $y > 1$, we have $y \leqslant -\frac{1}{2} + \sqrt{\frac{1}{4} + (x+1)^2}$, hence $\frac{1}{2} + \sqrt{\frac{1}{4} + x^2} < -\frac{1}{2} + \sqrt{\frac{1}{4} + (x+1)^2}$.

It is then easy to show that $\sqrt{\frac{1}{4} + x^2} < x$, which leads to a contradiction.

12. We prove the first statement by assuming the contrary. Since $a_1 \geqslant a_2 \geqslant \cdots \geqslant a_n$, $\max\{|a_1|, |a_2|, \cdots, |a_n|\} = a_1$ or $|a_n|$ (clearly $a_1 > 0$). But if $\max\{|a_1|, |a_2|, \cdots, |a_n|\} = |a_n|$, then $a_n < 0$, $|a_n| > a_1$.

We next set $a_1 \geqslant a_2 \geqslant \cdots \geqslant a_{n-k} > a_{n-k+1} = a_{n-k+2} = \cdots = a_n$.

Since $0 \leqslant \left|\frac{a_i}{a_n}\right| < 1$, there exists l, such that $\left|\frac{a_i}{a_n}\right|^{2l+1} < \frac{1}{n}$ $(1 \leqslant i \leqslant n-k)$, thus

$$\begin{aligned}
&\left(\frac{a_1}{a_n}\right)^{2l+1} + \left(\frac{a_2}{a_n}\right)^{2l+1} + \cdots + \left(\frac{a_n}{a_n}\right)^{2l+1} \\
&= \left(\frac{a_1}{a_n}\right)^{2l+1} + \left(\frac{a_2}{a_n}\right)^{2l+1} + \cdots + \left(\frac{a_{n-k}}{a_n}\right)^{2l+1} + k \\
&\geqslant k - \left|\frac{a_1}{a_n}\right|^{2l+1} - \left|\frac{a_2}{a_n}\right|^{2l+1} - \cdots - \left|\frac{a_{n-k}}{a_n}\right|^{2l+1} \\
&> k - \frac{n-k}{n} = k - 1 + \frac{k}{n} > 0.
\end{aligned}$$

As a result $a_1^{2l+1} + a_2^{2l+1} + \cdots + a_n^{2l+1} < 0$, which leads to a contradiction.

Now let us prove the second statement.

When $x > a_1$, $x - a_j > 0$, $1 \leqslant j \leqslant n$,

$$\begin{aligned}
(x-a_1)(x-a_2)\cdots(x-a_n) &\leqslant (x-a_1) \cdot \left[\frac{(x-a_2) + \cdots + (x-a_n)}{n-1}\right]^{n-1} \\
&= (x-a_1) \cdot \left(x - \frac{a_2 + a_3 + \cdots + a_n}{n-1}\right)^{n-1}.
\end{aligned}$$

Since $a_1+a_2+\cdots+a_n\geqslant 0$, $a_1\geqslant -(a_2+a_3+\cdots+a_n)$. Hence $x+\frac{1}{n-1}a_1\geqslant x-\frac{a_2+a_3+\cdots+a_n}{n-1}(>0)$, it follows that

$$(x-a_1)(x-a_2)\cdots(x-a_n)\leqslant (x-a_1)\left(x+\frac{a_1}{n-1}\right)^{n-1}$$
$$=(x-a_1)\cdot\sum_{s=0}^{n-1}C_{n-1}^s\left(\frac{a_1}{n-1}\right)^s\cdot x^{n-1-s}$$
$$=(x-a_1)\cdot\sum_{s=0}^{n-1}\frac{C_{n-1}^s}{(n-1)^s}\cdot a_1^s\cdot x^{n-1-s}.$$

It's easy to see that when $0\leqslant s\leqslant n-1$, $\frac{C_{n-1}^s}{(n-1)^s}\leqslant 1$, thus

$$(x-a_1)(x-a_2)\cdots(x-a_n)\leqslant (x-a_1)\cdot\sum_{s=0}^{n-1}a_1^s x^{n-1-s}\,(x>a_1\geqslant 0)$$
$$=(x-a_1)\cdot\frac{x^{n-1}-a_1^{n-1}\cdot\frac{a_1}{x}}{1-\frac{a_1}{x}}=x^n-a_1^n.$$

13. Assume to the contrary that there exists $1\leqslant i<j\leqslant n$ such that $A\geqslant 2a_ia_j$.

WLOG assume $i=1$, $j=2$, we thus have

$$A+\sum_{i=1}^n a_i^2\geqslant 2a_1a_2+\sum_{i=1}^n a_i^2=(a_1+a_2)^2+a_3^2+\cdots+a_n^2.$$

According to Cauchy's Inequality, $(a_1+a_2)^2+a_3^2+\cdots+a_n^2\geqslant\frac{1}{n-1}\cdot(a_1+a_2+\cdots+a_n)^2$. Therefore $A+\sum_{i=1}^n a_i^2\geqslant\frac{1}{n-1}\left(\sum_{i=1}^n a_i\right)^2$, which leads to a contradiction.

As a result, for any $1\leqslant i<j\leqslant n$, $A<2a_ia_j$.

14. We first show that $a_k-a_{k+1}\geqslant 0$ by assuming the contrary.

Assume there exists some $a_k<a_{k+1}$, then $a_{k+1}\leqslant a_k-a_{k+1}+a_{k+2}<a_{k+2}$, the sequence $\{a_s\mid s=k,k+1,\cdots\}$ is monotonically increasing.

Hence $\sum_{s=k}^{n}(n>k)$ tends to infinity when n tends to infinity, which leads to a contradiction. Thus $a_k - a_{k+1} \geqslant 0$, $k = 1, 2, \cdots$.

Denote $b_k = a_k - a_{k+1} \geqslant 0$, $k = 1, 2, \cdots$.

$$\begin{aligned} 1 &\geqslant a_1 + a_2 + \cdots + a_k \\ &= b_1 + 2b_2 + 3b_3 + \cdots + kb_k + a_{k+1} \\ &\geqslant (1 + 2 + 3 + \cdots + k)b_k = \frac{k(k+1)}{2}b_k. \end{aligned}$$

Therefore $b_k \leqslant \frac{2}{k(k+1)} < \frac{2}{k^2}$, and so $0 \leqslant a_k - a_{k+1} < \frac{2}{k^2}$.

15. Assume to the contrary that $ab > 0$, WLOG assume $a > 0$, thus $b > 0$. We discuss three possible cases in the following.

(1) If $c > 0$, all the roots of $x^4 + ax^3 + bx + c$ are negative, yet the coefficient of x^2 is 0, which is a contradiction.

(2) If $c < 0$, product of the four real roots is $c < 0$, so there will be 1 or 3 positive roots, and we divide this case further into two more cases.

(i) If there are 3 positive roots x_2, x_3, x_4 and 1 negative root x_1, then $x_1 + x_2 + x_3 + x_4 = -a < 0$, thus $-x_1 > -x_1 - a = x_2 + x_3 + x_4$. Since the coefficient of x^2 is 0, we must have $x_1(x_2 + x_3 + x_4) + x_2x_3 + x_2x_4 + x_3x_4 = 0$.

However, $x_2x_3 + x_2x_4 + x_3x_4 = -x_1(x_2 + x_3 + x_4) > (x_2 + x_3 + x_4)^2$, this leads to a contradiction.

(ii) If there are only 1 positive root x_1, and there are 3 negative roots x_2, x_3, x_4, we have $x_1x_2x_3x_4\left(\frac{1}{x_1} + \frac{1}{x_2} + \frac{1}{x_3} + \frac{1}{x_4}\right) = -b < 0$.

Since $x_1x_2x_3x_4 < 0$, $\frac{1}{x_1} + \frac{1}{x_2} + \frac{1}{x_3} + \frac{1}{x_4} > 0$. Also, note that $-x_1(x_2 + x_3 + x_4) = x_2x_3 + x_3x_4 + x_2x_4 > 0$, multiply these two inequalities and we obtain

$$\begin{aligned} -(x_2 + x_3 + x_4) &> \left(-\frac{1}{x_2} - \frac{1}{x_3} - \frac{1}{x_4}\right)(x_2x_3 + x_3x_4 + x_4x_2) \\ &= -2(x_2 + x_3 + x_4) - \left(\frac{x_2x_3}{x_4} + \frac{x_3x_4}{x_2} + \frac{x_4x_2}{x_3}\right), \end{aligned}$$

which leads to a contradiction.

(3) If $c = 0$, $x^3 + ax^2 + b = 0$ has 3 real roots x_1, x_2, x_3. Since $a > 0$, $b > 0$, all three roots are negative. But the coefficient in front of x is 0, so $x_1x_2 + x_2x_3 + x_3x_1 = 0$, which obviously cannot be true.

In sum, we have shown that $ab \leqslant 0$.

Exercise 5

1. Consider the identity $(k-a)(k-b)(k-c) = k^3 - (a+b+c)k^2 + (ab+bc+ca)k - abc$.

2. Consider the identity $(\alpha+\beta+\gamma-\alpha\beta\gamma)^2 = 2(\alpha\beta-1)(\beta\gamma-1)(\gamma\alpha-1) + 2 - \alpha^2\beta^2\gamma^2 + \alpha^2 + \beta^2 + \gamma^2$.

3. $|g(x)| = |cx^2 - c + bx + a + c| \leqslant |c| \cdot |x^2 - 1| + |bx + a + c| \leqslant 1 + |bx + a + c|$. Consider the function $T(x) = |bx + a + c|$, $-1 \leqslant x \leqslant 1$.

When $x = -1$, $T(-1) = |a - b + c| \leqslant 1$; when $x = 1$, $T(1) = |a + b + c| \leqslant 1$. Therefore when $|x| \leqslant 1$, $T(x) \leqslant 1$. Then $|g(x)| \leqslant 1 + T(x) \leqslant 2$, hence $F(x) \leqslant 2$, and the equality is reached when $x = \pm 1$.

4. Construct the function $f(x) = \dfrac{x}{1+x}$ it's easy to show that $f(x)$ is an increasing function on $[0, +\infty)$, thus $f(|a+b+c|) \leqslant f(|a|+|b|+|c|)$, or

$$\begin{aligned}\frac{|a+b+c|}{1+|a+b+c|} &\leqslant \frac{|a|+|b|+|c|}{1+|a|+|b|+|c|} \\ &= \frac{|a|}{1+|a|+|b|+|c|} + \frac{|b|}{1+|a|+|b|+|c|} \\ &\quad + \frac{|c|}{1+|a|+|b|+|c|} \\ &\leqslant \frac{|a|}{1+|a|} + \frac{|b|}{1+|b|} + \frac{|c|}{1+|c|}.\end{aligned}$$

So the conclusion follows.

5. Set $x = 2$, $y = z = 1$, we have $a \geqslant \frac{6}{5}$. Therefore, when $a \geqslant$

$\frac{6}{5}$, the original inequality has integer root (2, 1, 1), but (x, y, z) cannot form a triangle, hence $a < \frac{6}{5}$.

Also, since when $a < 1, x^2 + y^2 + z^2 \leqslant a(xy + yz + zx) < xy + yz + zx$, which leads to a contradiction. We conclude that $a \geqslant 1$.

Next, we show that when $1 \leqslant a < \frac{6}{5}$, any positive integer root for the original inequality can form three sides of a triangle.

First, (1, 1, 1) is a solution to the original inequality, and it forms a triangle. If the inequality has integer root (x_1, y_1, z_1) that cannot form a triangle, then transform the original inequality into $x^2 - a(y+z)x + y^2 + z^2 - ayz \leqslant 0$ (WLOG assume $x_1 \geqslant y_1 + z_1$).

Consider the function $f(x) = x^2 - a(y_1 + z_1)x + y_1^2 + z_1^2 - ay_1z_1$, its axis of symmetry is given by $x = \frac{a}{2}(y_1 + z_1) < \frac{3}{5}(y_1 + z_1) < y_1 + z_1 \leqslant x_1$, thus $(y_1 + z_1, y_1, z_1)$ is also a solution.

As a result, $a \geqslant \dfrac{(y_1 + z_1)^2 + y_1^2 + z_1^2}{(y_1 + z_1)y_1 + (y_1 + z_1)z_1 + y_1z_1} = u$, then $(u-2)y_1^2 + (3u-2)y_1z_1 + (u-2)z_1^2 = 0$.

Obviously $u \neq 2$, so $\Delta = (3u-2)^2 - 4(u-2)^2 \geqslant 0$, hence $u \geqslant \frac{6}{5}$, it follows that $a \geqslant u \geqslant \frac{6}{5}$, which leads to a contradiction.

6. WLOG assume that $\frac{b_1}{a_1} \geqslant \frac{b_2}{a_2} \geqslant \frac{b_3}{a_3}$. Consider the equation $(a_1a_2 + a_2a_3 + a_3a_1)x^2 - (a_1b_2 + a_2b_1 + a_3b_2 + a_2b_3 + a_3b_1 + a_1b_3)x + (b_1b_2 + b_2b_3 + b_3b_1) = (a_1x - b_1)(a_2x - b_2) + (a_2x - b_2)(a_3x - b_3) + (a_3x - b_3)(a_1x - b_1)$.

Since when $x = \frac{b_2}{a_2}$, the RHS of above $= \left(\frac{a_3b_2}{a_2} - b_3\right)\left(\frac{a_1b_2}{a_2} - b_1\right) \leqslant 0$. Consequently, this quadratic function on x must have real roots, so $\Delta \geqslant 0$ and the original inequality holds.

When the equality holds, $\Delta = 0$, so the above quadratic function on x must have equal roots $\frac{b_2}{a_2}$. Hence, either $\frac{b_1}{a_1} = \frac{b_2}{a_2}$ or $\frac{b_2}{a_2} = \frac{b_3}{a_3}$. WLOG assume that $\frac{b_1}{a_1} = \frac{b_2}{a_2} = \lambda$. We have

$$f(x) = (a_1a_2 + a_2a_3 + a_3a_1)\left(x - \frac{b_2}{a_2}\right)^2.$$

Therefore, $(a_1a_2 + a_2a_3 + a_3a_1)(x - \lambda)^2 = (a_1x - a_1\lambda)(a_2x - a_2\lambda) + (a_2x - a_2\lambda)(a_3x - b_3) + (a_3x - b_3)(a_1x - a_1\lambda)$.

It's now easy to show that $b_3 = a_3\lambda$, so the conclusion holds.

7. Let $f(x) = x^2 + \frac{1}{5}$, then as $x > 0$, $f(x)$ is monotonically increasing. Let $g(x) = f(x) - x = (x - r_1)(x - r_2)$, where $r_1 = \frac{5-\sqrt{5}}{10}$, $r_2 = \frac{5+\sqrt{5}}{10}$, $0 < r_1 < r_2 < 1$. Then $f(x) < x$, if $x \in (r_1, r_2)$. $f(x) \geqslant x$, if otherwise.

Fix $n \geqslant 5$, denote $a = |a_{n-5}|$, it suffices to show that $a_{n+5} \geqslant a_{n-4}^4$.

(1) $a \leqslant r_1$, then $a < 1$, $a_{n+5} \geqslant a > a^4 = a_{n-5}^4$. We have used the fact that $a \leqslant f(a) \leqslant f^{(2)}(a) \leqslant \cdots \leqslant f^{(10)}(a) = a_{n+5}$.

(2) $a \in [r_2, 1]$, we also have

$$a_{n+5} = f^{(10)}(a) > f^{(9)}(a) > \cdots > a > a^4 = a_{n-5}^4.$$

(3) $a \in (r_1, r_2)$, $r_1 = f(r_1) < f(r_2) = r_2$, $a_{n+5} = f^{(10)}(a) > r_1$, so it suffices to show that $r_1 > r_2^4$, which is easy to show.

(4) $a > 1$, then $a_{n+5} = f^{(10)}(a) > f^{(9)}(a) > \cdots > f^{(2)}(a) = \left(a^2 + \frac{1}{5}\right)^2 + \frac{1}{5} > a^4 = a_{n-5}^2$, so the conclusion follows.

Comment: We give two more proofs.

Proof 2: For any non-negative integer k, since $\left(a_k - \frac{1}{2}\right)^2 \geqslant 0$, we have

$$a_{k+1} + \frac{1}{20} \geqslant a_k^2 + \frac{1}{5} + \frac{1}{20} = a_k^2 + \frac{1}{4} \geqslant a_k.$$

Thus $a_{k+1} + \frac{1}{20} \geqslant a_k$.

For $k = n+1, n+2, n+3, n+4$, take the sum to get $a_{n+5} + \frac{1}{5} \geqslant a_{n+1} \geqslant a_n^2 + \frac{1}{5}$.

Similarly, we have $a_n \geqslant a_{n-5}^2$, thus $a_{n+5} \geqslant a_n^2 \geqslant a_{n-5}^4$, and the conclusion follows.

Proof 3: It suffices to show that $a_{n+5} \geqslant a_n^2 (n \geqslant 5)$.

Since $a_{n+5} \geqslant a_{n+4}^2 + \frac{1}{5}, \cdots, a_{n+1} \geqslant a_n^2 + \frac{1}{5}$, adding these yields

$$a_{n+5} \geqslant \sum_{i=1}^{4} (a_{n+i}^2 - a_{n+i}) + 1 + a_n^2 = \sum_{i=1}^{4} \left(a_{n+i} - \frac{1}{2}\right)^2 + a_n^2 \geqslant a_n^2.$$

Thus the conclusion holds.

8. Construct a rectangular coordinates system xOy, and consider three points $A(x, 0)$, $B\left(-\frac{y}{2}, -\frac{\sqrt{3}}{2}y\right)$, $C\left(\frac{z}{2}, \frac{\sqrt{3}}{2}z\right)$, the original inequality is equivalent to $|AB| + |AC| \geqslant |BC|$, which is obvious.

9. From the conditions we construct $\triangle ABC$, which contains a point P inside that satisfies $\angle APB = \angle BPC = \angle CPA = 120°$. Assume $PA = x$, $PB = y$, $PC = z$, then $AB = \sqrt{x^2 + xy + y^2}$, $BC = \sqrt{y^2 + yz + z^2}$, $CA = \sqrt{z^2 + zx + x^2}$.

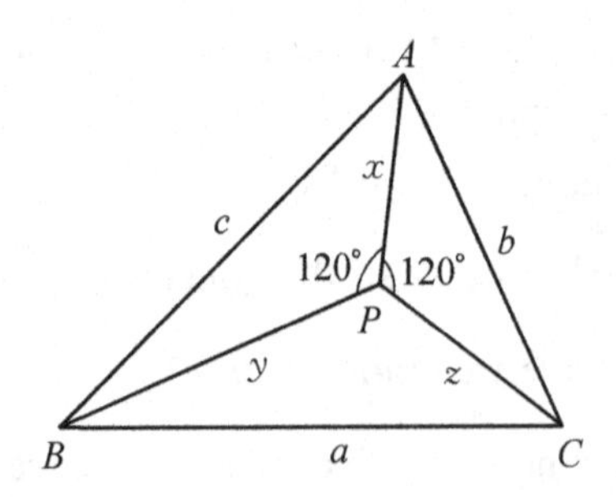

Since the sum of two sides of a triangle is larger than the third side, we have $\sqrt{x^2 + xy + y^2} + \sqrt{y^2 + yz + z^2} + \sqrt{z^2 + zx + x^2} \leqslant 2(x + y + z)$.

Also, since $a^2 + b^2 + c^2 \geqslant 4\sqrt{3}S$, $ab + bc + ca \geqslant 4\sqrt{3}S$, we have $(a +$

$b+c)^2 \geqslant 12\sqrt{3}S$, hence $a+b+c \geqslant 2\sqrt{3\sqrt{3}S} = 3\sqrt{xy+yz+zx}$.

10. From the given conditions, construct a rectangular cuboid $ABCD-A_1B_1C_1D_1$ with length, width and height of $\cos\alpha$, $\cos\beta$, $\cos\gamma$ alternatively, thus its diagonal must have length 1, as shown in the figure below ($AB = \cos\alpha$, $BC = \cos\beta$, $BB_1 = \cos\gamma$). It's easy to see that $\angle ABD_1 = \alpha$, $\angle CBB_1 = \beta$, $\angle B_1BD_1 = \gamma$.

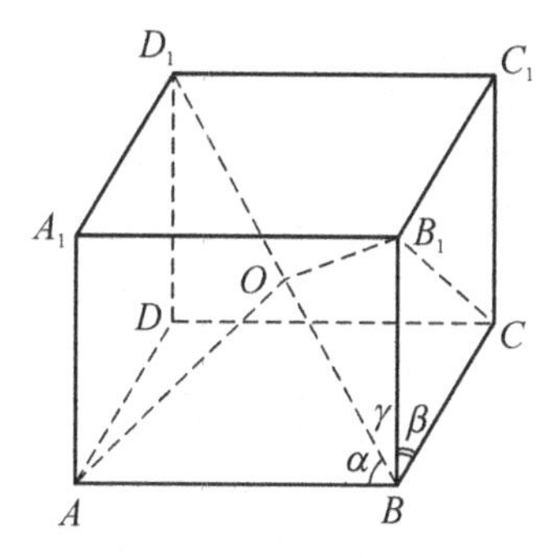

In the trihedral angle $B-AD_1C$, we have $\angle ABD_1 + \angle D_1BC > \angle ABC = \frac{\pi}{2}$, hence $\alpha+\beta > \frac{\pi}{2}$. Similarly, $\beta+\gamma > \frac{\pi}{2}$, $\gamma+\alpha > \frac{\pi}{2}$, yielding $\alpha+\beta+\gamma > \frac{3\pi}{4}$.

Pick the middle point O of BD_1, then $\angle AOD_1 = 2\alpha$, $\angle COD_1 = 2\beta$, $\angle BB_1D_1 = 2\gamma$. It's easy to show that $\angle COB_1 = \angle AOD_1 = 2\alpha$. Consider the trihedral angle $O-CB_1D_1$, we have $\angle COB_1 + \angle B_1OD_1 + \angle COD_1 < 2\pi$, thus $\alpha+\beta+\gamma < \pi$.

11. The original inequality is equivalent to

$$\sqrt{1-x^2} - \frac{\sqrt{2}-1}{2} \leqslant px+q \leqslant \sqrt{1-x^2} + \frac{\sqrt{2}-1}{2} \quad (0 \leqslant x \leqslant 1).$$

Using point $A\left(0, \frac{\sqrt{2}-1}{2}\right)$ and $B\left(0, -\frac{\sqrt{2}-1}{2}\right)$ as centers alternatively, and 1 as radius to construct circle A and B. Denote the arc of circle A in the first quadrant as l_1, the arc of circle B in the first and fourth quadrants as l_2. Thus when $0 \leqslant x \leqslant 1$, the line $y = px+q$ lies in between l_1 and l_2.

Since the line that connect the two end points of arc l_1 is actually tangent to circle B, it is tangent to arc l_2 as well. Therefore, the line $y = px + q$ must pass the two end points of l_1, so $p = -1$, $q = \frac{\sqrt{2}+1}{2}$.

12. If we consider $\sqrt{x^2+y^2}$ as the hypotenuse of a right-angle triangle, we can construct a graph in the following.

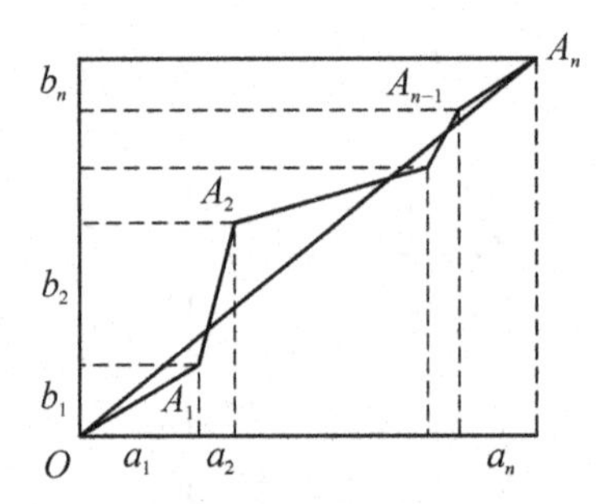

Hence, the LHS of the inequality $= OA_1 + A_1A_2 + \cdots + A_{n-1}A_n$, and the RHS of the inequality $= OA_n$. Since the length of the fold line $OA_1 \cdots A_n$ is no less than the straight line segment OA_n, $OA_1 + A_1A_2 + \cdots + A_{n-1}A_n \geqslant OA_n$, so the inequality holds.

13. (1) The original inequality is $a_1(a - a_2) + a_2(a - a_3) + a_3(a - a_4) + a_4(a - a_1) \leqslant 2a^2$, and we construct a square $ABCD$ with side length α in the following.

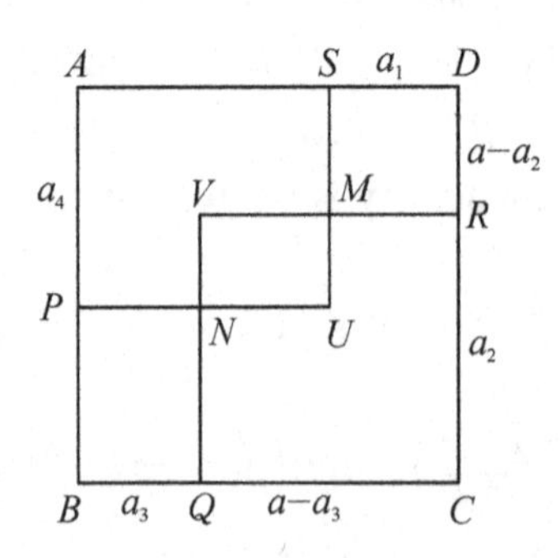

Pick U, V, such that $SD = a_1$, $CR = a_2$, $BQ = a_3$, $AP = a_4$. $SDRM$ and $CRVQ$ are non-overlapping, and $BPNQ$ and $APUS$ are non-overlapping, thus at most they cover the square $ABCD$ twice, so the original inequality holds.

(2) is similar to (1), and the six rectangles will cover the $a \times a$ at most three times.

14. WLOG assume that $x_1 \geqslant x_2 \geqslant \cdots \geqslant x_{100}$, we next construct a graph to show that $x_1 + x_2 + x_3 > 100$.

Consider x_i^2 as the area of a square with side length $x_i (1 \leqslant i \leqslant 100)$. Since $x_1 + x_2 + \cdots + x_{100} \leqslant 300$, all these squares can be placed one next to another inside a rectangle with length 300 and width 100, as shown in the graph below.

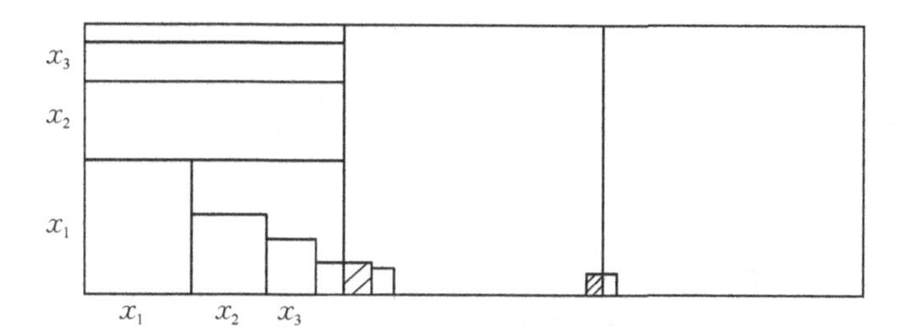

If $x_1 + x_2 + x_3 \leqslant 100$, the first square block of side length 100 contains three non-overlapping stripes with width x_1, x_2, x_3 alternatively.

Since x_i is monotonically decreasing, the second square block of side length 100 contains small squares (including incomplete ones as indicated as shaded area in the graph) with side length $x_i \leqslant x_3$. Clearly we can move these to the horizontal stripe with width x_2 in the first square block. Similarly, all the small squares contained in the third 100×100 square block can be placed inside the horizontal stripe with width x_3. As a result the sum of their areas $x_1^2 + x_2^2 + \cdots + x_{100}^2 < 100^2$, which leads to a contradiction.

Therefore $x_1 + x_2 + x_3 > 100$.

Comment: We give another proof in the following.

Proof 2: WLOG assume that $x_1 \geqslant x_2 \geqslant \cdots \geqslant x_{100}$. Denote $s = x_1 + x_2 + x_3$, $\lambda = x_3$. Let $\delta = x_2 - x_3 \geqslant 0$. Since $x_1 \geqslant x_2$, it's easy to show that $(x_1 + \delta)^2 + (x_2 - \delta)^2 \geqslant x_1^2 + x_2^2$, thus

$$x_1^2 + x_2^2 + x_3^2 \leqslant (s - 2\lambda)^2 + 2\lambda^2.$$

Since $\lambda \geqslant x_4 \geqslant x_5 \geqslant \cdots \geqslant x_{100}$, $x_4 + x_5 + \cdots + x_{100} < 300 - s$,

$x_4^2 + x_5^2 + \cdots + x_{100}^2 \leqslant \lambda(x_4 + x_5 + \cdots + x_{100}) \leqslant (300 - s)\lambda$.

Hence $x_1^2 + x_2^2 + \cdots + x_{100}^2 \leqslant 6\lambda^2 + (300 - 5s)\lambda + s^2$.

Denote $f(\lambda) = 6\lambda^2 + (300 - 5s)\lambda + s^2$, since $0 < \lambda \leqslant \frac{s}{3}$. Also, since $f(0) = s^2$, $f\left(\frac{s}{3}\right) = 100s$, we have $x_1^2 + x_2^2 + \cdots + x_{100}^2 \leqslant \max\{s^2, 100s\}$. In addition, from $x_1^2 + x_2^2 + \cdots + x_{100}^2 > 10\,000$, which implies $s = x_1 + x_2 + x_3 > 100$.

15. WLOG assume that $a_1 \leqslant a_2 \leqslant \cdots \leqslant a_n$. Then

$$A_3 = a_1^2 + a_2^2 + a_3^2 - a_1a_2 - a_2a_3 - a_3a_1 \geqslant 0.$$

$A_5 = (a_1 - a_2)(a_1 - a_3)(a_1 - a_4)(a_1 - a_5) + (a_2 - a_1)(a_2 - a_3)(a_2 - a_4)(a_2 - a_5) + (a_3 - a_1)(a_3 - a_2)(a_3 - a_4)(a_3 - a_5) + (a_4 - a_1)(a_4 - a_2)(a_4 - a_3)(a_4 - a_5) + (a_5 - a_1)(a_5 - a_2)(a_5 - a_3)(a_5 - a_4)$.

Since $(a_3 - a_1)(a_3 - a_2)(a_3 - a_4)(a_3 - a_5) \geqslant 0$, but the sum of the first 2 terms of A_5 is $(a_2 - a_1)[(a_3 - a_1)(a_4 - a_1)(a_5 - a_1) - (a_3 - a_2)(a_4 - a_2)(a_5 - a_2)] \geqslant 0$. Similarly, the sum of the last 2 terms of A_5 is also $\geqslant 0$, so $A_5 \geqslant 0$.

When $n \geqslant 7$, we construct the following counter example. Let $a_1 = a_2 = a_3 < a_4 < a_5 = a_6 = a_7 = \cdots = a_n$, then

$$A_n = (a_4 - a_1)(a_4 - a_2)(a_4 - a_3)(a_4 - a_5)(a_4 - a_6)\cdots(a_4 - a_n) < 0.$$

So the conclusion holds.

16. Construct the series $x_n = (1 + a_1)(1 + a_2)\cdots(1 + a_n) - (1 + a_1 + a_2 + \cdots + a_n)$ $(n \geqslant 2)$, then $x_{n+1} - x_n = a_{n+1}[(1 + a_1)\cdots(1 + a_n) - 1]$.

If $a_i > 0$ $(i = 1, 2, \cdots, n + 1)$, from the above it's easy to see that $x_{n+1} > x_n$. If $-1 < a_i < 0$ $(i = 1, 2, \cdots, n)$, then $0 < 1 + a_i < 1$, we also have $x_{n+1} > x_n$.

Therefore $\{x_n\}$ is a monotonically increasing series $(n \geqslant 2)$. Since $x_2 = (1 + a_1)(1 + a_2) - 1 - a_1 - a_2 = a_1a_2 > 0$, we conclude that the original inequality holds for any $n \geqslant 2$, $x_n > 0$.

17. Construct the series $x_n = 2C_n - D_n$, $y_n = D_n - C_n$, $n \in \mathbf{N}_+$, thus

$$
\begin{aligned}
x_{n+1} - x_n &= 2(C_{n+1} - C_n) - (D_{n+1} - D_n) \\
&= 2(a_{n+1} - b_{n+1})^2 - (a_{n+1} - b_{n+1})^2 - n(b_{n+1}^2 - b_n^2) \\
&\quad + 2(b_{n+1} - b_n)(a_1 + a_2 + \cdots + a_n) \\
&= (a_{n+1} - b_{n+1})^2 - n(b_{n+1}^2 - b_n^2) + 2nb_n(b_{n+1} - b_n) \\
&= [(n+1)b_{n+1} - nb_n - b_{n+1}]^2 - n(b_{n+1}^2 - b_n^2) + 2n(b_n b_{n+1} - b_n^2) \\
&= (n^2 - n)(b_{n+1} - b_n)^2 \geqslant 0.
\end{aligned}
$$

Also, since $x_1 = 2C_1 - D_1 = (a_1 - b_1)^2 = 0$, so for any n, $x_n \geqslant 0$ $(n \in \mathbf{N}_+)$. Similarly, $y_{n+1} - y_n = n(b_{n+1}^2 - b_n^2) - 2(b_{n+1} - b_n)(a_1 + a_2 + \cdots + a_n) = n(b_{n+1} - b_n)^2 \geqslant 0$. Finally, $y_1 = D_1 - C_1 = 0$, so for any $n \in \mathbf{N}_+$, $y_n \geqslant 0$.

As discussed above, $C_n \leqslant D_n \leqslant 2C_n$.

Exercise 6

1. It's easy to show that $\frac{1}{x_i} + \frac{1}{x_{i+1}} \geqslant \frac{4}{x_i + x_{i+1}}$ $(i = 1, 2, \cdots, n,$ $x_{n+1} = x_1)$, add these n inequalities together yields the desired result.

2. Since $a^3 + b^3 \geqslant a^2 b + b^2 a$, $\frac{1}{a^3 + b^3 + abc} \leqslant \frac{1}{ab(a+b+c)}$.

Similarly $\frac{1}{b^3 + c^3 + abc} \leqslant \frac{1}{bc(a+b+c)}$ and $\frac{1}{c^3 + a^3 + abc} \leqslant \frac{1}{ca(a+b+c)}$.

Add these three inequalities together yields the original inequality.

3. It's easy to show that $\frac{x}{1+y+zx} \leqslant \frac{1}{x+y+z}$, $\frac{y}{1+z+xy} \leqslant \frac{1}{x+y+z}$, and $\frac{z}{1+x+yz} \leqslant \frac{1}{x+y+z}$. Therefore the equality is achieved and we must have $x = y = z = 1$.

Comment: We can also show that $\frac{x}{1+y+zx} \leqslant \frac{x}{x+y+z}$ etc, which yields $x + y + z \geqslant 3$ and obtain $x = y = z = 1$.

4. Since $a^5+b^5-a^2b^2(a+b)=(a^2-b^2)(a^3-b^3)\geqslant 0$, $a^5+b^5\geqslant a^2b^2(a+b)$, therefore

$$\frac{ab}{a^5+b^5+ab}=\frac{ab\cdot abc}{a^5+b^5+ab\cdot abc}=\frac{a^2b^2c}{a^5+b^5+a^2b^2c}$$
$$\leqslant\frac{a^2b^2c}{a^2b^2(a+b)+a^2b^2c}=\frac{c}{a+b+c}.$$

Similarly, we have

$$\frac{bc}{b^5+c^5+bc}\leqslant\frac{a}{a+b+c},\ \frac{ca}{c^5+a^5+ca}\leqslant\frac{b}{a+b+c}.$$

Add these three inequalities together yields the original inequality, and the equality is achieved when $a=b=c=1$.

5. First, it's easy to show that $y^{2\alpha+3\beta}+z^{2\alpha+3\beta}\geqslant y^{\alpha+2\beta}z^{\alpha+\beta}+y^{\alpha+\beta}z^{\alpha+2\beta}$.

On the other hand, since $2004^{\alpha+\beta}+x^{\alpha}(y^{2\alpha+3\beta}+z^{2\alpha+3\beta})=x^{\alpha}[x^{\beta}y^{\alpha+\beta}z^{\alpha+\beta}+y^{2\alpha+3\beta}+z^{2\alpha+3\beta}]\geqslant x^{\alpha}[x^{\beta}y^{\alpha+\beta}z^{\alpha+\beta}+y^{\alpha+2\beta}z^{\alpha+\beta}+y^{\alpha+\beta}z^{\alpha+2\beta}]=x^{\alpha}y^{\alpha+\beta}z^{\alpha+\beta}(x^{\beta}+y^{\beta}+z^{\beta})$, we have

$$u\leqslant\frac{x^{\beta}}{2004^{\alpha+\beta}(x^{\beta}+y^{\beta}+z^{\beta})}+\frac{y^{\beta}}{2004^{\alpha+\beta}(x^{\beta}+y^{\beta}+z^{\beta})}$$
$$+\frac{z^{\beta}}{2004^{\alpha+\beta}(x^{\beta}+y^{\beta}+z^{\beta})}=\frac{1}{2004^{\alpha+\beta}},$$

thus $u_{\max}=2004^{-(\alpha+\beta)}$.

6. From the AM-GM Inequality,

$\frac{x_i^m}{a-x_i}+\frac{(a-x_i)a^{m-2}}{(n-1)^2n^{m-2}}+\frac{a^{m-1}}{(n-1)n^{m-1}}+\cdots+\frac{a^{m-1}}{(n-1)n^{m-1}}$ (repeating $m-2$ times) $\geqslant m\cdot\frac{x_i\cdot a^{m-2}}{(n-1)n^{m-2}}$.

Therefore,

$$\sum_{i=1}^{n}\frac{x_i^m}{a-x_i}\geqslant\sum_{i=1}^{n}\left[\frac{mx_ia^{m-2}}{(n-1)n^{m-2}}-\frac{(a-x_i)a^{m-2}}{(n-1)^2n^{m-2}}-\frac{(m-2)a^{m-1}}{(n-1)n^{m-1}}\right]$$
$$=\frac{a^{m-1}}{(n-1)n^{m-2}}.$$

Comment: We can also use Cauchy Inequality and Power Mean Inequality to solve this problem.

7. (1) $\sqrt[3]{\dfrac{a}{b+c}}=\dfrac{a}{\sqrt[3]{a\cdot a\cdot(b+c)}}\geqslant\dfrac{3a}{2a+b+c}>\dfrac{3a}{2a+2b+2c}$.

Similarly, we have $\sqrt[3]{\dfrac{b}{a+c}}>\dfrac{3b}{2a+2b+2c}$, $\sqrt[3]{\dfrac{c}{a+b}}>\dfrac{3c}{2a+2b+2c}$.

Add the above 3 inequalities yields the original inequality.

(2) $\sqrt[3]{\dfrac{a^2}{(b+c)^2}}=\dfrac{\sqrt[3]{2}\cdot a}{\sqrt[3]{2a(b+c)(b+c)}}\geqslant\dfrac{3\sqrt[3]{2}a}{2a+2b+2c}$.

Similarly, we have

$$\sqrt[3]{\frac{b^2}{(a+c)^2}}\geqslant\frac{3\sqrt[3]{2}b}{2a+2b+2c},\ \sqrt[3]{\frac{c^2}{(a+b)^2}}\geqslant\frac{3\sqrt[3]{2}c}{2a+2b+2c}.$$

Adding these together yields the original inequality (note that the equality can be achieved).

8. We generalize the condition to $v_1+v_2+\cdots+v_n=1$, $v_1\leqslant v_2\leqslant\cdots\leqslant v_n\leqslant rv_1$. Next, we show that

$$v_1^2+v_2^2+\cdots+v_n^2\leqslant\frac{(r+1)^2}{4m},\ r>1,\ r\in\mathbf{R}^+.$$

For any $j\in\mathbf{N}_+$, $1\leqslant j\leqslant n$, from the given condition we have $(v_j-v_1)(rv_1-v_j)\geqslant 0$, thus $rv_1v_j-rv_1^2-v_j^2+v_1v_j\geqslant 0$.

Sum the above from $j=1$ to n yields

$$\sum_{j=1}^{n}v_j^2\leqslant(r+1)v_1-rnv_1^2=-rn\left(v-1-\frac{r+1}{2rn}\right)^2+\frac{(r+1)^2}{4rn}\leqslant\frac{(r+1)^2}{4rn}.$$

9. (1) According to the AM-GM Inequality, we have

$\dfrac{a^5}{b^2-1}+\dfrac{25(5+\sqrt{15})}{12}(b-1)+\dfrac{25(5-\sqrt{15})}{12}(b+1)+\dfrac{25\sqrt{15}}{18}+\dfrac{25\sqrt{15}}{18}\geqslant\dfrac{125a}{6}$. Hence $\dfrac{a^5}{b^2-1}\geqslant\dfrac{125}{6}(a-b)+\dfrac{25}{18}\sqrt{15}$.

Similarly we have

$$\frac{b^5}{c^2-1}\geqslant\frac{125}{6}(b-c)+\frac{25}{18}\sqrt{15},\ \frac{c^5}{a^2-1}\geqslant\frac{125}{6}(c-a)+\frac{25}{18}\sqrt{15}.$$

Add these together yields the original inequality.

(2) Since $(b^3-1)(b^3-1)(b^3-1)\cdot\frac{3}{2}\cdot\frac{3}{2}\leqslant\frac{3^5\cdot b^{15}}{5^5}$, we have $b^3-1\leqslant\frac{3\sqrt[3]{20}\cdot b^5}{25}$, thus $\frac{a^5}{b^3-1}\geqslant\frac{5\sqrt[3]{50}\cdot a^5}{6b^5}$. Similarly,

$$\frac{b^5}{c^3-1}\geqslant\frac{5\sqrt[3]{50}\cdot b^5}{6c^5},\ \frac{c^5}{a^3-1}\geqslant\frac{5\sqrt[3]{50}\cdot c^5}{6a^5}.$$

So LHS of the inequality $\geqslant\frac{5\sqrt[3]{50}}{6}\left(\frac{a^5}{b^5}+\frac{b^5}{c^5}+\frac{c^5}{a^5}\right)\geqslant\frac{5}{2}\sqrt[3]{50}$.

Comment: We can use the method in (2) to solve (1). In fact, since

$$(b^2-1)(b^2-1)\cdot\frac{2}{3}\cdot\frac{2}{3}\cdot\frac{2}{3}\leqslant\left(\frac{2b^2}{5}\right)^5,$$

we have $b^2-1\leqslant\frac{6\sqrt{3}b^5}{25\sqrt{5}}$, so $\frac{a^5}{b^2-1}\geqslant\frac{25\sqrt{5}a^5}{6\sqrt{3}b^5}$ and it's then easy to show the original inequality.

10. From the given condition, $\frac{m_2}{M_1}\leqslant\frac{b_i}{a_i}\leqslant\frac{M_2}{m_1}$, $\frac{m_2}{M_1}a_i\leqslant b_i\leqslant\frac{M_2}{m_1}a_i$, hence $\left(b_i-\frac{M_2}{m_1}a_i\right)\left(b_i-\frac{m_2}{M_1}a_i\right)\leqslant 0$, thus

$$b_i^2-\left(\frac{M_2}{m_1}+\frac{m_2}{M_1}\right)a_ib_i+\frac{M_2m_2}{M_1m_1}a_i^2\leqslant 0.$$

According to the AM-GM Inequality,

$$2\left(\sum b_i^2\cdot\frac{M_2m_2}{M_1m_1}\sum a_i^2\right)^{1/2}\leqslant\sum b_i^2+\frac{M_2m_2}{M_1m_1}\sum a_i^2,$$

thus

$$2\left(\sum b_i^2\cdot\frac{M_2m_2}{M_1m_1}\sum a_i^2\right)^{1/2}\leqslant\left(\frac{M_2}{m_1}+\frac{m_2}{M_1}\right)\sum a_ib_i,$$

and therefore

$$\frac{\sum a_i^2\sum b_i^2}{(\sum a_ib_i)^2}\leqslant\frac{1}{4}\left(\frac{\frac{M_2}{m_1}+\frac{m_2}{M_1}}{\sqrt{\frac{M_2m_2}{M_1m_1}}}\right)^2=\frac{1}{4}\left(\sqrt{\frac{M_1M_2}{m_1m_2}}+\sqrt{\frac{m_1m_2}{M_1M_2}}\right)^2.$$

11. We can show that

$$x_i \cdot \sqrt{\frac{1-x_j}{x_j}} + x_j \cdot \sqrt{\frac{1-x_i}{x_i}} \geqslant (1-x_i)\sqrt{\frac{x_j}{1-x_j}} + (1-x_j)\sqrt{\frac{x_i}{1-x_i}},$$

and it's then easy to prove the original inequality.

12. Assume to the contrary that for any $i, j, t, l \in \{1, 2, \cdots, n\}$ (at least 3 are distinct to each other) we have $x_i - x_t \neq x_l - x_j$, we show next that for any k, $1 < k < n$,

$$\sum_{i=1}^{n} x_i > \frac{1}{k+1} \cdot \left[k \cdot \frac{n(n+1)(2n+1)}{6} - (k+1)^2 \frac{n(n+1)}{2}\right].$$

WLOG assume that $x_1 \leqslant x_2 \leqslant \cdots \leqslant x_n$.

From the assumption, $x_j - x_i$ are distinct to each other $(1 \leqslant i < j \leqslant m)$ and $x_i \geqslant 0$. Thus,

$$\begin{aligned} M &= \sum_{i=1}^{m-1}(x_{i+1} - x_i) + \sum_{i=1}^{m-2}(x_{i+2} - x_i) + \cdots + \sum_{i=1}^{m-k}(x_{i+k} - x_i) \\ &\geqslant 0 + 1 + 2 + 3 + \cdots + \left[\frac{(2m-1-k)k}{2} - 1\right] \\ &> \frac{1}{2}m^2k^2 - \frac{1}{2}k[k(k+1)+1]m. \end{aligned}$$

Also, note that $M = (x_m - x_1) + (x_m + x_{m-1} - x_2 - x_1) + \cdots + (x_m + x_{m-1} + \cdots + x_{m-k+1} - x_k - x_{k-1} - \cdots - x_1) \leqslant kx_m + (k-1)x_{m-1} + \cdots + x_{m-k-1} \leqslant kx_m + (k-1)x_m + \cdots + x_m = \frac{k(k+1)}{2}x_m$.

We have $x_m > \frac{k}{k+1}m^2 - \frac{k(k+1)+1}{k+1}m \geqslant \frac{k}{k+1}m^2 - (k+1)m$ $(k < m \leqslant n)$, which also holds when $1 \leqslant m \leqslant k$.

As a result, $\sum_{i=1}^{n} x_i > \frac{k}{k+1} \sum_{i=1}^{n} i^2 - (k+1) \sum_{i=1}^{n} i = \frac{1}{k+1} \cdot \left[k \cdot \frac{n(n+1)(2n+1)}{6} - (k+1)^2 \cdot \frac{n(n+1)}{2}\right]$ holds for any $1 < k < n$.

Exercise 7

1. Use mathematical induction to show that $a_n > n$.

2. Assume $(k-1)(a_1 + a_3 + \cdots + a_{2k-1}) \geqslant k(a_2 + a_4 + \cdots + a_{2k-2})$. To prove the statement for n, it suffices to show that $a_1 + a_2 + \cdots + a_{2k-1} + ka_{2k+2} \geqslant a_2 + a_4 + \cdots + a_{2k-2} + (k+1)a_{2k}$, or $k(a_{2k+1} - a_{2k}) \geqslant (a_2 - a_1) + \cdots + (a_{2k} - a_{2k-1})$. From the given condition that $a_{i+2} - a_{i+1} \geqslant a_{i+1} - a_i$, this is quite obvious.

3. Use the fact that $y = -x^2 + x$ is monotonic on $\left(0, \frac{1}{4}\right)$ during the induction step from k to $k+1$.

4. Since $a_n > 0$, arc cot a_{n+1} + arccot $a_{n+2} \in (0, \pi)$. cot x is decreasing on the interval $(0, \pi)$, so it suffices to show that cot (arc cot a_n) $\geqslant$ cot(arc cot a_{n+1} + arc cot a_{n+2}), or $(a_{n+1} + a_{n+2})a_n \geqslant a_{n+1} \cdot a_{n+2} - 1$.

But $a_{n+2} = a_{n+1} + a_n$, so it suffices to show that $a_n a_{n+2} - a_{n+1}^2 \geqslant -1$. Using mathematical induction, it's easy to show that $a_n a_{n+2} - a_{n+1}^2 = (-1)^{n+1}$, so the original inequality holds, and the equal sign is achieved if and only if n is even.

5. We use the induction on n. When $n = 1$, $a_1 \geqslant a_1$, the inequality obviously holds.

Assume that the inequality holds for $n = 1, 2, \cdots, k-1$, that is

$$a_1 \geqslant a_1,$$

$$a_1 + \frac{a_2}{2} \geqslant a_2,$$

$$\cdots\cdots$$

$$a_1 + \frac{a_2}{2} + \cdots + \frac{a_{k-1}}{k-1} \geqslant a_{k-1}.$$

Adding these together, we have $(k-1)a_1 + (k-2)\frac{a_2}{2} + \cdots + (k-(k-1))\frac{a_{k-1}}{k-1} \geqslant a_1 + a_2 + \cdots + a_{k-1}$, that is, $k\left(a_1 + \frac{a_2}{2} + \cdots + \frac{a_{k-1}}{k-1}\right) \geqslant 2(a_1 + a_2 + \cdots + a_{k-1}) = (a_1 + a_{k-1}) + (a_2 + a_{k-2}) + \cdots + (a_{k-1} + a_1) \geqslant$

$ka_k - a_k$.

Therefore, $a_1 + \frac{a_2}{2} + \cdots + \frac{a_{k-1}}{k-1} \geqslant a_k$, so the original inequality holds.

6. If there exists x, such that $a_n = [nx]$, then $\frac{a_n}{n} \leqslant x < \frac{a_n + 1}{n}$. This inequality should hold for $n = 1, 2, \cdots, 2004$. Hence, x has to be inside 2004 intervals characterized by $\left[\frac{a_n}{n}, \frac{a_n + 1}{n}\right)$. Therefore if x exists, we can set x as $\max\left\{\frac{a_m}{m}\right\}$.

As such, it suffices to show that for any $n \in \{1, 2, \cdots, 2004\}$, $\frac{a_n + 1}{n} > x \geqslant \frac{a_m}{m}$.

We show in the following that if m, n are positive integers, and $m, n \leqslant 2004$,

$$ma_n + m > na_m. \qquad ①$$

When $m = n$, ① holds. Also, when $m = 1$, $n = 2$ and $m = 2$, $n = 1$, we have $a_2 + 1 > 2a_1$ and $2a_1 + 2 > a_2$, so ① holds.

Assume that the statement is true for $m, n < k (3 \leqslant k \leqslant 2004)$.

When $m = k$, assume $m = nq + r$, $q \in \mathbf{N}_+$, $0 \leqslant r < n$. Then $a_m = a_{nq+r} \leqslant a_{nq} + a_r + 1 \leqslant a_{(q-1)n} + a_n + a_r + 2 \leqslant \cdots \leqslant qa_n + a_r + q$. Hence $na_m \leqslant n(qa_n + a_r + q) = ma_n + m + (na_r - ra_n - r)$. From the induction assumption, $ra_n + r > na_r$, so $na_m < ma_n + m$.

When $n = k$, assume $n = mq + r$, $q \in \mathbf{N}_+$, $0 \leqslant r < m$, thus $a_n \geqslant a_{qm} + a_r \geqslant \cdots \geqslant qa_m + a_r$. $ma_n + m \geqslant mqa_m + ma_r + m = na_m + ma_r + m - ra_m$. By the induction assumption, $ma_r + m > ra_n$, hence $ma_n + m > na_m$, so ① holds, and $\frac{a_n + 1}{n} > \frac{a_m}{m}$. Let $x = \max\limits_{1 \leqslant m < 2004}\left(\frac{a_m}{m}\right)$, thus $\frac{a_n + 1}{n} > x$, and so $nx - 1 < a_n \leqslant nx$, therefore $a_n = [nx]$.

7. Let $y_n = x_n - \sqrt{2}$, we use induction to show that $|y_n| < \left(\frac{1}{2}\right)^n$, $n \geqslant 3$.

Assume that $|y_n| < \left(\frac{1}{2}\right)^n$, from the given conditions, $y_{n+1} + \sqrt{2} = 1 + y_n + \sqrt{2} - \frac{1}{2}(y_n + \sqrt{2})^2$, thus $|y_{n+1}| = \left|\frac{1}{2}y_n(2 - 2\sqrt{2} - y_n)\right| = \frac{1}{2}|y_n| \cdot |y_n + 2\sqrt{2} - 2| < \frac{1}{2} \cdot \left(\frac{1}{2}\right)^n \cdot \left|\left(\frac{1}{2}\right)^n + 2\sqrt{2} - 2\right| < \frac{1}{2} \cdot \left(\frac{1}{2}\right)^n \cdot 1 = \left(\frac{1}{2}\right)^{n+1}$, and the conclusion follows.

8. We use induction to prove that when $1 \leqslant k \leqslant n$,

$$\sqrt{ka + \sqrt{(k+1)a + \cdots + \sqrt{na}}} < 1 + \sqrt{ka}. \quad ①$$

When $k = n$, ① is obviously true. Now assume that

$$\sqrt{(k+1)a + \sqrt{(k+2)a + \cdots + \sqrt{na}}} < 1 + \sqrt{(k+1)a},$$

then

$$\begin{aligned} \sqrt{ka + \sqrt{(k+1)a + \cdots + \sqrt{na}}} &< \sqrt{ka + 1 + \sqrt{(k+1)a}} \\ &< \sqrt{ka + 1 + 2\sqrt{ka}} = 1 + \sqrt{ka}. \end{aligned}$$

Hence ① holds for any k, $1 \leqslant k \leqslant n$. In particular, $k = 1$ yields the original inequality.

9. When $n = 1$, $a_1 = 1$, the conclusion holds. Assume that when $n = k$ the conclusion holds. When $n = k + 1$, $a_1 a_2 \cdots a_{k+1} = 1$, from the induction assumption, we have

$$\begin{gathered} a_{k+1} \cdot a_1 + a_2 + a_3 + \cdots + a_n \geqslant k, \\ a_{k+1} \cdot a_2 + a_1 + a_3 + \cdots + a_n \geqslant k, \\ \cdots\cdots \\ a_{k+1} \cdot a_k + a_1 + a_2 + \cdots + a_{k-1} \geqslant k. \end{gathered}$$

Adding these together, $(a_1 + a_2 + \cdots + a_k)(a_{k+1} + k - 1) \geqslant k^2$. Therefore

$$\sum_{i=1}^{k+1} a_i = \sum_{i=1}^{k} a_i + a_{k+1} \geqslant \frac{k^2}{a_{k+1} + k - 1} + a_{k+1} \geqslant k + 1,$$

so the conclusion follows.

10. Assume $b_k = \frac{a_k}{2^k}$, then $b_{k+2} = \frac{1}{2}b_{k+1} + \frac{1}{4}b_k$, $b_1 = \frac{1}{2}$, $b_2 = \frac{1}{4}$, thus

$$b_{k+1} = \frac{1}{2}b_k + \frac{1}{4}b_{k-1},\ b_k = \frac{1}{2}b_{k-1} + \frac{1}{4}b_{k-2},\ \cdots,\ b_3 = \frac{1}{2}b_2 + \frac{1}{4}b_1,$$

adding these together yields $S_{k+2} - b_1 - b_2 = \frac{1}{2}S_{k+1} - \frac{1}{2}b_1 + \frac{1}{4}S_k$, $S_{k+2} = \frac{1}{2}S_{k+1} + \frac{1}{4}S_k + \frac{1}{2}$, and via induction it's easy to show that $S_n < 2$.

11. When $n = 1$, the inequality obviously holds. We use induction to show that when $n = 2^k$, the inequality holds.

For $k = 1$, we have

$$\frac{1}{r_1+1} + \frac{1}{r_2+1} - \frac{2}{\sqrt{r_1 r_2}+1} = \frac{(\sqrt{r_1 r_2}-1)(\sqrt{r_1}-\sqrt{r_2})^2}{(r_1+1)(r_2+1)(\sqrt{r_1 r_2}+1)} \geqslant 0.$$

If for $k = m$, the inequality holds, consider the case when $k = m+1$. That is, we need to show that if the original inequality holds for n numbers, it also holds for $2n$ numbers.

If $r_1, r_2, \cdots, r_{2n}$ are all greater than 1, we have

$$\begin{aligned}\sum_{i=1}^{2n} \frac{1}{r_i+1} &= \sum_{i=1}^{n} \frac{1}{r_i+1} + \sum_{i=n+1}^{2n} \frac{1}{r_i+1} \\ &\geqslant \frac{n}{\sqrt[n]{r_1 r_2 \cdots r_n}+1} + \frac{n}{\sqrt[n]{r_{n+1}\cdots r_{2n}}+1} \\ &\geqslant \frac{2n}{\sqrt[2n]{r_1 r_2 \cdots r_{2n}}+1},\end{aligned}$$

so the original inequality holds.

For any positive integer n, there exists positive integer k, such that $m = 2^k > n$.

Assume $r_{n+1} = r_{n+2} = \cdots = r_m = \sqrt[n]{r_1 r_2 \cdots r_n}$, then

$$\frac{1}{r_1+1} + \frac{1}{r_2+1} + \cdots + \frac{1}{r_n+1} + \frac{m-n}{\sqrt[n]{r_1 r_2 \cdots r_n}+1} \geqslant \frac{m}{\sqrt[n]{r_1 r_2 \cdots r_n}+1},$$

so the original inequality holds.

12. From the given conditions,

$$\frac{1}{f(n)}=\frac{1}{f(n)-1}-\frac{1}{f(n+1)-1},$$

hence

$$\sum_{k=1}^{n}\frac{1}{f(k)}=\frac{1}{f(1)-1}-\frac{1}{f(n+1)-1}=1-\frac{1}{f(n+1)-1}.$$

Next, we show that $2^{2^{n-1}}<f(n+1)-1<2^{2^n}$, which leads to the conclusion.

When $n=1$, $f(2)=3$, so $2<f(2)<4$. Assume that when $n=m$, $2^{2^{m-1}}<f(m+1)-1<2^{2^m}$. Consider the case when $n=m+1$. Since $f(m+2)=f(m+1)(f(m+1)-1)+1$, we have

$$(2^{2^{m-1}}+2)(2^{2^{m-1}}+1)+1\leqslant f(m+2)\leqslant 2^{2^m}(2^{2^m}-1)+1.$$

Therefore $2^{2^m}<f(m+2)-1<2^{2^{m+1}}$.

13. First let us estimate the upper bound for $\{f(n)\}$. Due to the monotonicity of f, it suffices to investigate its subsequence $\{f(2^k)\}$. Now, from the given condition, for $k\in\mathbf{N}_+$, we have

$$\begin{aligned}f(2^{k+1})&\leqslant\left(1+\frac{1}{2^k}\right)f(2^k)\\&\leqslant\left(1+\frac{1}{2^k}\right)\left(1+\frac{1}{2^{k-1}}\right)f(2^{k-1})\\&\leqslant\cdots\leqslant 2\lambda_k,\end{aligned}$$

where $\lambda_k=(1+1)\left(1+\frac{1}{2}\right)\cdots\left(1+\frac{1}{2^k}\right)$.

We can guess $\lambda_k<5$ from some simple calculations. In fact, we can use induction to prove a stronger statement that $\lambda_k\leqslant 5\left(1-\frac{1}{2^k}\right)<5$. For any $f\in S$ and any $n\in\mathbf{N}_+$ there exists a positive integer k, such that $n<2^{k+1}$, then $f(n)\leqslant f(2^{k+1})\leqslant 2\lambda_k<10$.

In the next step, we create a function f_0 satisfying the conditions and that it achieves a value >9 somewhere. Note that $2\lambda_5>9$, we construct $f_0 : \mathbf{N}_+ \to \mathbf{R}$ in the following way:

$$f_0(1) = 2,\ f_0(n) = 2\lambda_k (2^k < n \leqslant 2^{k+1}),\ k \in \{0, 1, 2, \cdots\}.$$

For any positive integer n, it's easy to see that $f_0(n+1) \geqslant f_0(n)$. We next show that $f_0(n) \geqslant \frac{n}{n+1} f_0(2n)$.

Assume $k \in \mathbf{N}$, such that $2^k < n \leqslant 2^{k+1}$, then $2^{k+1} < 2n \leqslant 2^{k+2}$. As a result,

$$f_0(2n) = 2\lambda_{k+1} = \left(1 + \frac{1}{2^{k+1}}\right) \cdot 2\lambda_k \leqslant \left(1 + \frac{1}{n}\right) f_0(n).$$

Finally, $f_0(2^6) = 2\lambda_5 > 9$, so we have solved the problem, and $M = 10$.

14. We apply the induction on n. When $n = 3$, since the original inequality is cyclic symmetrical in a_1, a_2, a_3, WLOG assume that a_1 is the largest. If $a_2 < a_3$, then $a_1^3 a_2 + a_2^3 a_3 + a_3^3 a_1 - (a_1 a_2^3 + a_2 a_3^3 + a_3 a_1^3) = (a_1 - a_2)(a_2 - a_3)(a_1 - a_3)(a_1 + a_2 + a_3) < 0$.

In consequence, we only need to consider the case when $a_2 \geqslant a_3$, which implies $a_1 \geqslant a_2 \geqslant a_3$. Hence, $a_1^3 a_2 + a_2^3 a_3 + a_3^3 a_1 \leqslant a_1^3 a_2 + 2a_1^2 a_2 a_3 \leqslant a_1^3 a_2 + 3a_1^2 a_2 a_3 = a_1^2 a_2 (a_1 + 3a_3) = \frac{1}{2} a_1 \cdot a_1 \cdot 3a_2 \cdot (a_1 + 3a_3) \leqslant \frac{1}{3} \cdot \left(\frac{a_1 + a_1 + 3a_2 + a_1 + 3a_3}{4}\right)^4 = \frac{1}{3} \cdot \left[\frac{3(a_1 + a_2 + a_3)}{4}\right]^4 = 27$, and the equality holds when $a_1 = 3$, $a_2 = 1$, $a_3 = 0$.

Assume when $n = k$ $(k \geqslant 3)$, the original inequality holds.

When $n = k + 1$, assume again that a_1 is the largest. Then

$$\begin{aligned} & a_1^3 a_2 + a_2^3 a_3 + \cdots + a_k^3 a_{k+1} + a_{k+1}^3 a_1 \\ \leqslant & a_1^3 a_2 + a_1^3 a_3 + (a_2 + a_3)^3 a_4 + \cdots + a_{k+1}^3 a_1 \\ = & a_1^3 (a_2 + a_3) + (a_2 + a_3)^3 a_4 + \cdots + a_{k+1}^3 a_1. \end{aligned}$$

Note that the k numbers a_1, $a_2 + a_3$, a_4, $\cdots$, a_{k+1} satisfy the condition

$$a_1 + (a_2 + a_3) + \cdots + a_{k+1} = 4.$$

From the induction assumption, $a_1^3(a_2 + a_3) + (a_2 + a_3)^3 a_4 + \cdots +$

$a_{k+1}^3 a_1 \leqslant 27$, therefore the original inequality holds when $n = k+1$.

15. We use the induction on n. When $n = 1$, the conclusion obviously holds. Assume when $n = s$, we have $\sum_{k=1}^{s} \frac{x^{k^2}}{k} \geqslant x^{s(s+1)/2}$. Consider the case when $n = s+1$.

$$\sum_{k=1}^{s+1} \frac{x^{k^2}}{k} = \sum_{k=1}^{s} \frac{x^{k^2}}{k} + \frac{x^{(s+1)^2}}{s+1} \geqslant x^{s(s+1)/2} + \frac{x^{(s+1)^2}}{s+1},$$

so it suffices to show that $x^{s(s+1)/2} + \frac{x^{(s+1)^2}}{s+1} \geqslant x^{(s+1)(s+2)/2}$, or

$$1 + \frac{x^{(s+1)(s+2)/2}}{s+1} \geqslant x^{s+1}. \quad ①$$

Denote $y = x^{(s+1)/2}$, $y > 0$, ① is equivalent to

$$1 + \frac{y^{s+2}}{s+1} \geqslant y^2. \quad ②$$

Let $f(y) = 1 + y^2\left(\frac{y^s}{s+1} - 1\right)$, where $y > 0$.

If $y \geqslant (s+1)^{1/s}$, then $f(y) \geqslant 1 > 0$, ② obviously holds.

If $y \in (0, 1]$, ② also obviously holds.

Now for $y \in (1, (s+1)^{1/s})$, we need to prove $f(y) > 0$. In this case, $\frac{y^s}{s+1} < 1$, thus $f(y) = 1 - y^2\left(1 - \frac{y^s}{s+1}\right)$.

For some undetermined coefficient A, we have

$$\begin{aligned} A^2\left[y^2\left(1 - \frac{y^s}{s+1}\right)\right]^s &= (Ay^s)(Ay^s)\left(1 - \frac{y^s}{s+1}\right)\cdots\left(1 - \frac{y^s}{s+1}\right) \\ &\leqslant \left\{\frac{1}{s+2}\left[2Ay^s + s\left(1 - \frac{y^s}{s+1}\right)\right]\right\}^{s+2}. \end{aligned}$$

Set $A = \frac{s}{2(s+1)}$, we have

$$\left(\frac{s}{2(s+1)}\right)^2 \cdot \left[y^2\left(1 - \frac{y^s}{s+1}\right)\right]^s \leqslant \left(\frac{s}{s+2}\right)^{s+2}.$$

Hence $y^2\left(1-\frac{y^s}{s+1}\right)\leqslant\frac{s}{s+2}\left[\frac{2(s+1)}{s+2}\right]^{2/s}$. When $s=1$, we have $y^2\left(1-\frac{y}{2}\right)\leqslant\frac{16}{27}<1$, and $f(y)>0$.

Finally, it suffices to show that for positive integer $s\geqslant 2$, $\left[\frac{2(s+1)}{s+2}\right]^{2/s}<\frac{s+2}{s}$.

This is equivalent to $\frac{2(s+1)}{s+2}<\left(1+\frac{2}{s}\right)^{s/2}$, and thus it suffices to show that $\left(1+\frac{2}{s}\right)^{s/2}\geqslant 2$.

In fact,

$$\left(1+\frac{2}{s}\right)^{s/2}=1+s\cdot\frac{2}{s}+C_s^2\cdot\left(\frac{2}{s}\right)^2+\cdots\geqslant 1+2+\frac{2(s-1)}{s}\geqslant 4,$$

so the conclusion holds.

16. We use induction on n. When $n=1$ the conclusion obviously holds. Assume the original inequality holds for $n-1$, that is

$$\sum_{1\leqslant i<j\leqslant n-1}|s_j-z_i|\leqslant\sum_{k=1}^{n-1}((n-k)|z_k|+(k-2)|s_k|).$$

But $\sum_{1\leqslant i<j\leqslant n}|s_j-z_i|=\sum_{1\leqslant i<j\leqslant n-1}|s_j-z_i|+\sum_{k=1}^{n}|s_n-z_k|$, and $\sum_{k=1}^{n}((n+1-k)|z_k|+(k-2)|s_k|)=\sum_{k=1}^{n-1}[(n-k)|z_k|+(k-2)|s_k|]+\sum_{k=1}^{n}|z_k|+(n-2)|s_n|$.

To prove the inequality for n, it suffices to show that

$$\sum_{k=1}^{n}|s_n-z_k|\leqslant\sum_{k=1}^{n}|z_k|+(n-2)|s_n|. \qquad ①$$

When $n=1, 2$, ① obviously holds. Next we show that ① holds for $n=3$. In other words,

$$|z_1+z_2|+|z_2+z_3|+|z_3+z_1|$$
$$\leqslant|z_1|+|z_2|+|z_3|+|z_1+z_2+z_3|.$$

Since $\sum_{k=1}^{3} | s_3 - z_k | \leqslant \sum_{k=1}^{3} | z_k | + | s_3 |$, taking the square on each side yields

$$| s_3 |^2 = \sum_{k=1}^{3} | z_k |^2 + \sum_{1 \leqslant i < j \leqslant 3} (z_i \bar{z}_j + \bar{z}_i z_j).$$

However, $\sum_{k=1}^{3} | s_3 - z_k |^2 = | s_3 |^2 + \sum_{k=1}^{3} | z_k |^2$, thus

$$\begin{aligned}
&\text{RHS}^2 - \text{LHS}^2 \\
&= 2 \sum_{1 \leqslant i < j \leqslant 3} | z_i z_j | + 2 \sum_{i=1}^{3} | s_3 z_i | - 2 \sum_{1 \leqslant i < j \leqslant 3} | s_3^2 - s_3 z_i - s_3 z_j + z_i z_j | \\
&\geqslant 2 \sum_{i=1}^{3} | s_3 z_i | - 2 \sum_{1 \leqslant i < j \leqslant 3} | s_3^2 - s_3 z_i - s_3 z_j | \\
&= 2 \sum_{i=1}^{3} | s_3 z_i | - 2 \sum_{1 \leqslant i < j \leqslant 3} | s_3 (s_3 - z_i - z_j) | = 0.
\end{aligned}$$

Assume that ① holds for $n \leqslant k$. When $n = k + 1$, consider $\sum_{i=1}^{k} z_i$ as 1 number, and use the results when $n = 3$, we have

$$\begin{aligned}
&| s_{k+1} - z_{k+1} | + | s_{k+1} - z_k | + \left| s_{k+1} - \sum_{i=1}^{k-1} z_i \right| \\
&\leqslant \left| \sum_{i=1}^{k-1} z_i \right| + | z_k | + | z_{k+1} | + | s_{k+1} |.
\end{aligned} \quad ②$$

Now, regard $z_k + z_{k+1}$ as a number, from the induction assumption,

$$\begin{aligned}
&\sum_{i=1}^{k-1} | s_{k+1} - z_i | + \left| \sum_{i=1}^{k-1} z_i \right| \\
&\leqslant | z_k + z_{k+1} | + \sum_{i=1}^{k-1} | z_i | + (k-2) | s_{k+1} |.
\end{aligned} \quad ③$$

Adding ② and ③, we have

$$\sum_{i=1}^{k+1} | s_{k+1} - z_i | \leqslant \sum_{i=1}^{k+1} | z_i | + (k + 1 - 2) | s_{k+1} |,$$

so the conclusion holds for $n = k + 1$, and we have completed the induction steps.

Exercise 8

1. Fix z, first find the minimum and maximum of $2x^2 + y$, then vary z to find that $S_{\max} = 3$, $S_{\min} = \frac{57}{72}$.

2. WLOG assume that C is an acute angle. Since $0 \leqslant \lambda_i \leqslant 2$, $0 < \frac{1}{2}(\lambda_1 A + \lambda_2 B) < \pi$, $0 < \frac{1}{6}(\lambda_1 A + \lambda_2 B + 4\lambda_3 C) < \pi$, $0 < \frac{1}{3}(\lambda_1 A + \lambda_2 B + \lambda_3 C) < \pi$. Hence $\sin \frac{1}{2}(\lambda_1 A + \lambda_2 B) > 0$, $\sin \frac{1}{6}(\lambda_1 A + \lambda_2 B + 4\lambda_3 C) > 0$, $\sin \frac{1}{3}(\lambda_1 A + \lambda_2 B + \lambda_3 C) > 0$.

As a result,

$$\begin{aligned}
&\sin \lambda_1 A + \sin \lambda_2 B + \sin \lambda_3 C + \sin \frac{\lambda_1 A + \lambda_2 B + \lambda_3 C}{3} \\
&= 2\sin \frac{\lambda_1 A + \lambda_2 B}{2} \cdot \cos \frac{\lambda_1 A - \lambda_2 B}{2} + 2\sin \frac{\lambda_1 A + \lambda_2 B + 4\lambda_3 C}{6} \cdot \\
&\quad \cos \frac{2\lambda_3 C - \lambda_1 A - \lambda_2 B}{6} \\
&\leqslant 2\sin \frac{\lambda_1 A + \lambda_2 B}{2} + 2\sin \frac{\lambda_1 A + \lambda_2 B + 4\lambda_3 C}{6} \\
&= 4\sin \frac{4\lambda_1 A + 4\lambda_2 B + 4\lambda_3 C}{12} \cdot \cos \frac{2\lambda_1 A + 2\lambda_2 B - 4\lambda_3 C}{12} \\
&\leqslant 4\sin \frac{\lambda_1 A + \lambda_2 B + \lambda_3 C}{3}.
\end{aligned}$$

The equality holds if and only if $\lambda_1 A = \lambda_2 B = \lambda_3 C$.

3. First consider fixing $\theta_1 + \theta_2$, we have $\sin^2\theta_1 + \sin^2\theta_2 = (\sin \theta_1 + \sin \theta_2)^2 - 2 \sin \theta_1 \sin \theta_2 = 4\sin^2 \frac{\theta_1 + \theta_2}{2} \cos^2 \frac{\theta_1 - \theta_2}{2} - \cos(\theta_1 - \theta_2) + \cos(\theta_1 + \theta_2) = 2\cos^2 \frac{\theta_1 - \theta_2}{2} \cdot \left(2\sin^2 \frac{\theta_1 + \theta_2}{2} - 1\right) + 1 + \cos(\theta_1 + \theta_2)$.

Hence, when $\theta_1 + \theta_2 < \frac{\pi}{2}$, and one of θ_1 and θ_2 is 0, the above reaches its maximum; when $\theta_1 + \theta_2 > \frac{\pi}{2}$, the larger the absolute difference between the two angles, the larger the above formula.

When $n \geqslant 4$, we can always find two angles whose sum is $\leqslant \frac{\pi}{2}$, we can adjust one of these two angles to 0, and the other to the sum of the original two angles so that the sum of squared sin's does not change. Therefore, it suffices to consider the case of 3 angles.

When $n = 3$, if two among θ_1, θ_2, θ_3 are $\frac{\pi}{2}$, and the other is 0, we can adjust these 3 angles to $\frac{\pi}{2}, \frac{\pi}{4}, \frac{\pi}{4}$. Assume that $\theta_1 \leqslant \theta_2 \leqslant \theta_3$, $\theta_1 < \theta_3$, then $\theta_1 + \theta_3 > \frac{\pi}{2}$, $\theta_1 < \frac{\pi}{3} < \theta_3$.

Let $\theta'_1 = \frac{\pi}{3}$, $\theta'_2 = \theta_2$, $\theta'_3 = \theta_1 + \theta_3 - \theta'_1$, then $\theta'_1 + \theta'_3 = \theta_1 + \theta_3$, $|\theta'_1 - \theta'_3| < |\theta_1 - \theta_3|$, from the previous discussions, we have $\sin^2\theta_1 + \sin^2\theta_2 + \sin^2\theta_3 < \sin^2\theta'_1 + \sin^2\theta'_2 + \sin^2\theta'_3$. Also, since $\theta'_2 + \theta'_3 = \frac{2\pi}{3}$, $\sin^2\theta'_1 + \sin^2\theta'_2 \leqslant 2\sin^2\frac{\pi}{3} = \frac{3}{2}$, therefore $\sin^2\theta_1 + \sin^2\theta_2 + \sin^2\theta_3 \leqslant \frac{9}{4}$, and the equality holds if and only if $\theta_1 = \theta_2 = \theta_3 = \frac{\pi}{3}$. We conclude that the maximum value when $n \geqslant 3$ is $\frac{9}{4}$.

Finally, when $n = 2$, $\theta_1 + \theta_2 = \pi$, so $\sin^2\theta_1 + \sin^2\theta_2 = 2\sin^2\theta_1 \leqslant 2$, so the maximum value is 2.

4. When one of a, b, c, d is 0, WLOG assume that $d = 0$, it suffices to show that $abc \leqslant \frac{1}{2}(ab + bc + ca)$. Since $\frac{1}{a} + \frac{1}{b} + \frac{1}{c} + a + b + c \geqslant 6$, we have $\frac{1}{a} + \frac{1}{b} + \frac{1}{c} \geqslant 2$, so the conclusion holds.

If none of a, b, c, d is 0, it suffices to show that

$$f = \sum \frac{1}{a} - \frac{1}{2}\sum \frac{1}{ab} \leqslant 1.$$

WLOG assume that $a \leqslant 1 \leqslant b$, let $a' = 1$, $b' = a + b - 1$, we have $a'b' \geqslant ab$, hence

$$f(a, b, c, d) - f(a', b', c', d')$$
$$= (a + b)\left(\frac{1}{ab} - \frac{1}{a'b'}\right)\left[1 - \frac{1}{2}\left(\frac{1}{a+b} + \frac{1}{c} + \frac{1}{d}\right)\right]$$
$$\leqslant (a + b)\frac{a'b' - ab}{a'b'ab}\left(1 - \frac{1}{2} \cdot \frac{32}{4}\right) \leqslant 0.$$

Thus $f(a, b, c, d) \leqslant f(a', b', c', d') = f(1, a + b - 1, c, d)$. Going through another two rounds of mollification, we obtain

$$\begin{aligned} f(1, a + b - 1, c, d) &\leqslant f(1, 1, a + b + c - 2, d) \\ &\leqslant f(1, 1, 1, a + b + c + d - 3) \\ &= f(1, 1, 1, 1) = 1. \end{aligned}$$

5. $S = (x_1 + x_2 + x_3 + x_4)^2 - \sum\limits_{i=1}^{4} x_i^3 - 6 \sum\limits_{1 \leqslant i < j < k \leqslant 4} x_i x_j x_k$
$= 3x_1^2(1 - x_1) + 3x_2^2(1 - x_2) + 3x_3^2(1 - x_3) + 3x_4^2(1 - x_4)$.

Thus $S \geqslant 0$, and the equality is achieved when one of $x_i (1 \leqslant i \leqslant 4)$ is 1 and the other three are 0.

Also, note that $\frac{1}{3}S = x_1^2(1 - x_1) + x_2^2(1 - x_2) + x_3^2(1 - x_3) + x_4^2(1 - x_4)$. One of $x_1 + x_2$ and $x_3 + x_4$ must be $\leqslant \frac{2}{3}$, WLOG assume that $x_1 + x_2 \leqslant \frac{2}{3}$.

We adjust $S = S(x_1, x_2, x_3, x_4)$ to $S' = S(x_1 + x_2, 0, x_3, x_4)$. It's easy to show that $S' \geqslant S$. Denote $S' = S(x'_1, x'_2, x'_3, 0)$. Then $x'_1 + x'_2 + x'_3 = 1$. So one of x'_1, x'_2, x'_3 must be $\geqslant \frac{1}{3}$, WLOG assume that $x'_3 \geqslant \frac{1}{3}$, so $x'_1 + x'_2 \leqslant \frac{2}{3}$.

Now, adjust $S' = S(x'_1, x'_2, x'_3, 0)$ to $S'' = S(x'_1 + x'_2, x'_3, 0, 0)$. Similarly it's easy to show that $S'' \geqslant S'$, where $S'' = S(a, b, 0, 0)$ and $a + b = 1$.

Therefore $S'' = a^2(1 - a) + b^2(1 - b) = ab \leqslant \frac{1}{4}$, so $S \leqslant 3S'' \leqslant \frac{3}{4}$.

The equality is achieved when $x_1 = \frac{1}{2}$, $x_2 = x_3 = 0$, $x_4 = \frac{1}{2}$. We conclude that $S \in \left[0, \frac{3}{4}\right]$.

6. Let $y_i = \sum_{j=i}^{n} x_j$, then $y_1 = n$, $\sum_{i=1}^{n} y_i = 2n - 2$, thus $S = \sum_{k=1}^{n} k^2 x_k$

$$= \sum_{k=1}^{n} k^2 (y_k - y_{k+1}) + n^2 y_n = \sum_{k=1}^{n} (2k-1) y_k.$$

Since $y_1 = n$, $y_2 \geqslant y_3 \geqslant \cdots \geqslant y_n$, if there exists $i \in \{2, 3, \cdots, n-1\}$, such that $y_i > y_n$ (denote i the largest index such that $y_i > y_n$). Because

$$(2i-1)y_i + (2i+1+\cdots+2n-1)y_n$$
$$< (2i-1+\cdots+2n-1) \cdot \frac{y_i + (n-i)y_n}{n-i+1} \text{(this is equivalent to } y_i > y_n\text{)},$$

it is clear that, keeping $y_i + y_{i+1} + \cdots + y_n$ unchanged, we can use their average to replace each of them and increase S. Therefore, when $y_2 = y_3 = \cdots = y_n$, $S_{\max} = n^2 - 2$.

7. When $n = 1$, $\sum_{j=1}^{n} (x_j^4 - x_j^5) = 0$.

When $n = 2$, $\sum_{j=1}^{n} (x_j^4 - x_j^5) = (x_1^4 + x_2^4) - (x_1 + x_2)(x_1^4 - x_1^3 x_2 + x_1^2 x_2^2 - x_1 x_2^3 + x_2^4) = x_1^3 x_2 - x_1^2 x_2^2 + x_1 x_2^3 = x_1 x_2 (1 - 3x_1 x_2)$.

Since $x_1 x_2 \leqslant \frac{1}{4}$, we find the maximum value is $\frac{1}{12}$.

When $n \geqslant 3$, we have

$$\begin{aligned}
&-(x^4 + y^4 - x^5 - y^5) + [(x+y)^4 - (x+y)^5] \\
&= [(x+y)^4 - x^4 - y^4] - [(x+y)^5 - x^5 - y^5] \\
&= xy(4x^2 + 4y^2 + 6xy) - xy(5x^3 + 5y^3 + 10x^2 y + 10xy^2) \\
&\geqslant xy(4x^2 + 4y^2 + 6xy) - 5xy(x+y)^3 \\
&= xy\left(\frac{7}{2}x^2 + \frac{7}{2}y^2 + \frac{x^2 + y^2}{2} + 6xy\right) - 5xy(x+y)^3 \\
&\geqslant xy\left(\frac{7}{2}x^2 + \frac{7}{2}y^2 + 7xy\right) - 5xy(x+y)^3 > 0.
\end{aligned}$$

This is equivalent to $\frac{7}{2}xy(x+y)^2 \cdot [7-10(x+y)] > 0$, or $x+y < \frac{7}{10}$.

When $n \geqslant 3$, there must be two terms with sum $< \frac{2}{3} < \frac{7}{10}$, keep replacing x, y with $x+y$ until finally we reach the case $n = 2$.

In conclusion, the maximum value of $\sum_{j=1}^{n}(x_j^4 - x_j^5)$ is $\frac{1}{12}$.

8. Denote $f(a_1, a_2, a_3) = (a_1 + a_2 + a_3 + 3\sqrt[3]{a_1a_2a_3}) - 2(\sqrt{a_1a_2} + \sqrt{a_2a_3} + \sqrt{a_3a_1})$.

WLOG assume that $a_1 \leqslant a_2 \leqslant a_3$, and we make the adjustment according to the following. Let

$$a'_1 = a_1, \; a'_2 = a'_3 = \sqrt{a_2a_3} = A,$$

then $f(a'_1, a'_2, a'_3) = (a_1 + 2A + 3a_1^{1/3}A^{2/3}) - 2[A + 2(a_1A)^{1/2}] = a_1 + 3a_1^{1/3}A^{2/3} - 4(a_1A)^{1/2} \geqslant 4\sqrt[4]{a_1(a_1^{1/3})^3(A^{2/3})^3} - 4(a_1A)^{1/2} = 0$.

We prove next that $f(a_1, a_2, a_3) \geqslant f(a'_1, a'_2, a'_3)$.

In fact, $f(a_1, a_2, a_3) - f(a'_1, a'_2, a'_3) = a_2 + a_3 + 4\sqrt{a_1A} - 2(\sqrt{a_1a_2} + \sqrt{a_2a_3} + \sqrt{a_3a_1}) = a_2 + a_3 - 2a_1^{1/2}(a_2^{1/2} + a_3^{1/2} - 2A^{1/2}) - 2A$.

Note that $a_1 \leqslant A$, $a_2^{1/2} + a_3^{1/2} \geqslant 2\sqrt[4]{a_2a_3} = 2A^{1/2}$. Therefore $f(a_1, a_2, a_3) - f(a'_1, a'_2, a'_3) \geqslant a_2 + a_3 - 2\sqrt{A}(\sqrt{a_2} + \sqrt{a_3} - 2\sqrt{A}) - 2A = (\sqrt{a_2} - \sqrt{A})^2 + (\sqrt{a_3} - \sqrt{A})^2 \geqslant 0$, so the original inequality holds.

Comment: Using the same method, it's easy to show the Kouber Inequality:

If $a_i \geqslant 0$, $n \geqslant 2$, the numbers in the sequence $a = (a_1, a_2, \cdots, a_n)$ are distinct to each other, then

$$(n-2)\sum_{i=1}^{n} a_i + n(a_1a_2\cdots a_n)^{1/n} - 2\sum_{1 \leqslant i < j \leqslant n}(a_ia_j)^{1/2} \geqslant 0.$$

9. We go through two steps to solve this problem:

(1) First consider when we have determined $a_1, a_2, \cdots, a_{10}$, how

to arrange them to minimize the corresponding sum. Consider the simple case first where $a_1+a_2+\cdots+a_{10}=55$, $\{a_1, a_2, \cdots, a_{10}\}=\{1, 2, \cdots, 10\}$, and we place $a_1, a_2, \cdots, a_{10}$ along the circle in order. We show that the two sides of 10 must be placed by 1 and 2.

If 1 is not a neighbor to 10, then we flip the part $a_i, a_{i+1}, \cdots, 1$, denote $S=a_1a_2+a_2a_3+\cdots+a_{10}a_1$ and S' to be the sum after the flip. Thus

$$S'-S=10\cdot 1+a_i\cdot a_j-(10\cdot a_i+a_j\cdot 1)\leqslant 0.$$

As a result the sum does not increase after adjustment. Similarly, we can show that the other side of 10 must be 2.

Using the same method, we can show that the other side of 1 must be 9, so on and so forth. It turns out that the only cyclic arrangement that minimizes S is 10, 1, 9, 3, 7, 5, 6, 4, 8, 2, which yields $S_{\min}=224$.

(2) Next we consider the case when $a_1+a_2+\cdots+a_{10}=2002$. Assume when $a_1+a_2+\cdots+a_{10}=n$, $\min S(n)=g(n)$, then $g(55)=224$.

Next we show that $g(n+1)\geqslant g(n)+3$.

Consider a cyclic arrangement $a_1, a_2, \cdots, a_{10}$ such that $a_1+a_2+\cdots+a_{10}=n+1$. If $a_1, a_2, \cdots, a_{10}$ are 10 adjacent positive integers, assume $a_1>1$ (if $a_1=1$, then $n=55$), so a_1-1 is not among the original 10 numbers, and replacing a_1 by a_1-1 reduces the sum by $a_1(1+2)-(a_1-1)(1+2)=3$.

However, if $a_1, a_2, \cdots, a_{10}$ are not adjacent positive integers, there must be one a_i such that a_i-1 is not among the original 10 numbers, so replacing a_i by a_i-1 also reduces the sum by 3. Therefore, for any arrangement $S(n+1)$ for $a_1+a_2+\cdots+a_{10}=n+1$, we can find an arrangement of $a_1, a_2, \cdots, a_i-1, \cdots, a_{10}$ such that $S(n+1)\geqslant S(n)+3\geqslant g(n)+3$, thus $\min S(n+1)\geqslant g(n)+3$ so $g(n+1)\geqslant g(n)+3$, and it follows that $g(2002)\geqslant 6065$. The equality is reached when $a_1=1$, $a_2=9$, $a_3=3$, $a_4=7$, $a_5=5$, $a_6=6$, $a_7=4$, $a_8=8$, $a_9=2$, $a_{10}=1957$.

10. Denote $f(a, b, c) = \frac{1}{a+b} + \frac{1}{b+c} + \frac{1}{c+a}$. WLOG assume that $a \leqslant b \leqslant c$. We first show that $f(0, a+b, c') \leqslant f(a, b, c)$, where $c' = \frac{1}{a+b}$, $ab + bc + ca = 1$.

In fact, $f(0, a+b, c') \leqslant f(a, b, c)$ is equivalent to

$$\frac{1}{a+b} + \frac{1}{a+b+c'} + \frac{1}{c'} \leqslant \frac{1}{a+b} + \frac{1}{b+c} + \frac{1}{c+a}.$$

Since $c' = \frac{1}{a+b}$, $\frac{1-ab}{a+b} = c$, the above inequality can be simplified as

$$(a+b)^2 ab \leqslant 2(1-ab).$$

Note that $2(1-ab) = 2c(a+b) \geqslant \frac{2(a+b)^2}{2} \geqslant (a+b)^2 ab$, so the conclusion follows. As such,

$$f(a, b, c) \geqslant \frac{1}{a+b} + \frac{1}{a+b+\frac{1}{a+b}} + a + b,$$

and it's easy to show that f reaches its minimum value $\frac{5}{2}$ when $a = 0$, $b = c = 1$.

Comment: We can also use the iterated method to find the extremum.

Clearly, if a, b, c are all $> \frac{\sqrt{3}}{3}$, we have $ab + bc + ca = 1$, which does not satisfy the conditions. Hence one of a, b, c must not exceed $\frac{\sqrt{3}}{3}$, WLOG assume $b \leqslant \frac{\sqrt{3}}{3}$. Then

$$\begin{aligned} S &= \frac{1}{a+b} + \frac{1}{b+c} + \frac{1}{c+a} = \frac{a+2b+c}{1+b^2} + \frac{1}{c+a} \\ &= \frac{2b}{1+b^2} + \frac{a+c}{1+b^2} + \frac{1}{a+c}. \end{aligned}$$

In the next step, we fix b and find the minimum value of $\frac{a+c}{1+b^2} +$

$\frac{1}{a+c}$.

Let $x = a+c$, so $f(x) = \frac{x}{1+b^2} + \frac{1}{x}$ is monotonically increasing when $x \geqslant \sqrt{1+b^2}$, but $1-b(a+c) = ac \leqslant \frac{(a+c)^2}{4}$, thus $\frac{x^2}{4} \geqslant 1 - bx$, which implies $x \geqslant 2\sqrt{b^2+1} - 2b$. Also, $b^2+1 \geqslant 4b^2$, hence $x \geqslant \sqrt{b^2+1}$, and $f(x) \geqslant f(2\sqrt{b^2+1}-2b)$.

From the above discussion,

$$S \geqslant \frac{2b}{1+b^2} + f(2\sqrt{b^2+1}-2b) = \frac{2}{\sqrt{b^2+1}} + \frac{\sqrt{b^2+1}+b}{2}.$$

Let $y = \sqrt{b^2+1} - b > 0$, then $b = \frac{1-y^2}{2y}$. Thus, $S \geqslant \frac{2}{y+b} + \frac{1}{2y} = \frac{9y^2+1}{2y(1+y^2)} = \frac{5}{2} + \frac{(1-y)\left(5\left(y-\frac{2}{5}\right)^2+\frac{1}{5}\right)}{2y(1+y^2)} \geqslant \frac{5}{2}$ (note that $\sqrt{b^2+1} - b \leqslant 1$).

The equality is achieved when $b = 0$, and $a = c = 1$, so the minimum value is $\frac{5}{2}$.

11. First, we calculate the maximum of $f(a_1, a_2, \cdots, a_{2001}) = a_1a_2 + a_2a_3 + \cdots + a_{2000}a_{2001} + a_{2001}a_1$.

Let us show that there exists $i \in \{1, 2, \cdots, 2001\}$ such that $a_i > a_{i+4}$ (denote $a_{2001+i} = a_i$). Assume the contrary, then $a_1 \leqslant a_5 \leqslant a_9 \leqslant \cdots \leqslant a_{2001} \leqslant a_4 \leqslant a_8 \leqslant \cdots \leqslant a_{2000} \leqslant a_3 \leqslant a_7 \leqslant \cdots \leqslant a_{1999} \leqslant a_2 \leqslant a_6 \leqslant \cdots \leqslant a_{1998} \leqslant a_1$.

This implies $a_1 = a_2 = \cdots = a_{2001} = \frac{2}{2001}$, which does not satisfy the given conditions.

Now WLOG assume $a_{1998} > a_1$. For given $a_1, a_2, \cdots, a_{1998}, a_{2000}$, let $a'_1 = a_1$, $a'_2 = a_2$, $\cdots$, $a'_{1998} = a_{1998}$, $a'_{2000} = a_{2000}$, $a'_{2001} = 0$, $a'_{1999} = a_{1999} + a_{2001}$. Then $\sum_i a'_i = 2$.

$f(a'_1, a'_2, \cdots, a'_{2001}) = a_1a_2 + a_2a_3 + \cdots + a_{1997}a_{1998} + a_{1998}(a_{1999} + a_{2001}) +$

$(a_{1999}+a_{2001})a_{2000} > f(a_1, a_2, \cdots, a_{2001})$.

Next, let $a_1''=a_1'$, $\cdots$, $a_{1998}''=a_{1998}'+a_{2000}'$, $a_{1999}''=a_{1999}'$, $a_{2000}'=0$, then it's easy to show that $f(a_1'', a_2'', \cdots, a_{2000}'', 0) > f(a_1', a_2', \cdots, a_{2000}', 0)$.

In fact, each that we can adjust the last non-zero a_i to 0, and turn the sum of a_i and a_{i-2} to the new a_{i-2}, f will increase till there are only 3 numbers left. In the end, $a_1+a_2+a_3=2$, $f(a_1, a_2, a_3, 0, \cdots, 0)=a_1a_2+a_2a_3=a_2(2-a_2)\leqslant 1$, so the maximum of $f(a_1, a_2, \cdots, a_{2001})$ is 1, and is achieved when $a_4=a_5=\cdots=a_{2001}=0$, $a_2=1$.

Therefore, $S=a_1^2+a_2^2+a_3^2=1+a_1^2+a_3^2\geqslant 1+\dfrac{(a_1+a_3)^2}{2}=\dfrac{3}{2}$. In addition, $S=2(a_1-1)^2+\dfrac{3}{2}\leqslant 2(0\leqslant a_1\leqslant 1)$, so $S_{\max}=2$, $S_{\min}=\dfrac{3}{2}$.

12. $\sum\limits_{1\leqslant i<j\leqslant n} x_ix_j(x_i+x_j)=\dfrac{1}{2}\sum\limits_{1\leqslant i<j\leqslant n} x_ix_j(x_i+x_j)+\dfrac{1}{2}\sum\limits_{1\leqslant i<j\leqslant n} x_jx_i(x_j+x_i)=\dfrac{1}{2}\sum\limits_{i\neq j}x_ix_j(x_i+x_j)=\dfrac{1}{2}\sum\limits_{i\neq j}x_i^2x_j+\dfrac{1}{2}\sum\limits_{i\neq j}x_ix_j^2=\dfrac{1}{2}\sum\limits_{i=1}^{n}x_i^2(1-x_i)+\dfrac{1}{2}\sum\limits_{j=1}^{n}x_j^2(1-x_j)=\sum\limits_{i=1}^{n}x_i^2-\sum\limits_{i=1}^{n}x_i^3$.

When $n=1$, $x_1=1$, the RHS of the above is 0.

When $n=2$, the RHS is $(x_1^2+x_2^2)-(x_1^3+x_2^3)$.

$x_1x_2\leqslant\dfrac{1}{4}(x_1+x_2)^2=\dfrac{1}{4}$, and the equality holds when $x_1=x_2=\dfrac{1}{2}$.

When $n\geqslant 3$, WLOG assume $x_1\geqslant x_2\geqslant\cdots\geqslant x_{n-1}\geqslant x_n$. Denote $v=\{(x_1, x_2, \cdots, x_n)\mid x_i\geqslant 0, 1\leqslant i\leqslant n, \sum\limits_{i=1}^{n}x_i=1\}$ and write

$$F(v)=\sum_{1\leqslant i<j\leqslant n}x_ix_j(x_i+x_j)=\sum_{i=1}^{n}x_i^2-\sum_{i=1}^{n}x_i^3.$$

Set $w=\{(x_1, x_2, \cdots, x_{n-2}, x_{n-1}+x_n, 0)\}$, then

$$F(w)-F(v)=x_{n-1}x_n[2-3(x_{n-1}+x_n)].$$

Also, since $\dfrac{1}{2}(x_{n-1}+x_n)\leqslant\dfrac{1}{n}(x_1+x_2+\cdots+x_n)$, $x_{n-1}+x_n\leqslant\dfrac{2}{n}\leqslant\dfrac{2}{3}$, so $F(w)\geqslant F(v)$.

We then rearrange $x_1, x_2, \cdots, x_{n-2}, x_{n-1}+x_n$ from largest to smallest as $x'_1, x'_2, \cdots, x'_{n-1}$, and obtain $F(v) \leqslant F(x'_1, x'_2, \cdots, x'_{n-1}, 0)$.

From previous discussion, $F(v) \leqslant F(x'_1, x'_2, \cdots, x'_{n-1}, 0) \leqslant F(x''_1, x''_2, \cdots, x''_{n-2}, 0, 0)$, and $\sum_{k=1}^{n-2} x''_k = 1$.

Finally, $F(v) \leqslant F(a, b, 0, \cdots, 0) = a^2+b^2-a^3-b^3 \leqslant \frac{1}{4}$ (where $a+b=1$, $a, b \in [0, 1]$). The equality holds when $x_1 = x_2 = \frac{1}{2}$, $x_3 = x_4 = \cdots = x_n = 0$.

13. WLOG assume that $x_1 \geqslant x_2 \geqslant \cdots \geqslant x_n \geqslant 0$, and the last non-zero number in $x_1, x_2, \cdots, x_n$ is x_{k+1} $(k \geqslant 2)$. Adjust $x_1, x_2, \cdots, x_k, x_{k+1}, 0, 0, \cdots, 0$ ($n-k-1$ zero's) to $x_1, x_2, \cdots, x_{k-1}, x_k+x_{k+1}, 0, 0, \cdots, 0$ ($n-k$ zero's) whose sum is still 1.

Denote $F(x_1, x_2, \cdots, x_k, x_{k+1}, \cdots, x_n) = \sum_{1 \leqslant i < j \leqslant n} x_i x_j (x_i^2 + x_j^2)$, after the adjustment $F(x_1, x_2, \cdots, x_{k-1}, x_k+x_{k+1}, 0, 0, \cdots, 0) - F(x_1, x_2, \cdots, x_k, x_{k+1}, 0, 0, \cdots, 0) = x_k x_{k+1} \cdot [(x_k+x_{k+1})(3-4(x_k+x_{k+1}))+2x_k x_{k+1}]$.

Since $k \geqslant 2$, $1 \geqslant x_1 + x_k + x_{k+1}$. As $x_1 \geqslant x_2 \geqslant \cdots \geqslant x_k \geqslant x_{k+1}$, we have $x_1 \geqslant \frac{x_k + x_{k+1}}{2}$, so $1 \geqslant \frac{3}{2}(x_k + x_{k+1})$, or $x_k + x_{k+1} \leqslant \frac{2}{3}$, thus $4(x_k + x_{k+1}) \leqslant \frac{8}{3} < 3$.

Therefore, $F(x_1, x_2, \cdots, x_{k-1}, x_k+x_{k+1}, 0, 0, \cdots, 0) \geqslant F(x_1, x_2, \cdots, x_k, x_{k+1}, 0, 0, \cdots, 0)$.

After several steps of adjustment, $F(x_1, x_2, \cdots, x_n) \leqslant F(a, b, 0, 0, \cdots, 0) = ab(1-2ab) \leqslant \frac{1}{8}$ (where $a \geqslant 0$, $b \geqslant 0$, and $a+b=1$).

Thus $c = \frac{1}{8}$, and the equality holds when two positive numbers in a_i are equal and the rest are zeros.

Comment: We show a simpler proof in the following.

Let $x_1 = x_2 = \frac{1}{2}$, and the other $x_i = 0$ $(3 \leqslant i \leqslant n)$, we have $c \geqslant$

$\frac{1}{8}$. We show next that

$$\left(\sum_{i=1}^{n} x_i\right)^4 \geqslant 8 \sum_{1 \leqslant i < j \leqslant n} x_i x_j (x_i^2 + x_j^2).$$

In fact,

$$\begin{aligned}\left(\sum_{i=1}^{n} x_i\right)^4 &= \left(\sum_{i=1}^{n} x_i^2 + 2\sum_{i<j} x_i x_j\right)^2 \\ &\geqslant 4\sum_{i=1}^{n} x_i^2 \cdot 2\sum_{i<j} x_i x_j = 8 \sum_{1 \leqslant i<j \leqslant n} \left(x_i x_j \cdot \sum_{i=1}^{n} x_i^2\right) \\ &\geqslant 8 \sum_{1 \leqslant i<j \leqslant n} x_i x_j (x_i^2 + x_j^2).\end{aligned}$$

14. Assume $l = 1997x$, $m = 1997y$, $n = 1997z$, then l, m, $n \in \{0, 1, 2, \cdots, 1996\}$, and $l^2 + m^2 - n^2 = 1997^2$. Next we first find the minimum value of $l + m + n$. Fix l, we have $(m - n)(m + n) = (1997 - l)(1997 + l)$. In order to make $m + n$ as small as possible, we have to let $m + n$ and $m - n$ as close as possible, but we always have $m + n \geqslant \sqrt{(1997 + l)(1997 - l)}$. Assume $l_1 = 1997 - l$, then

$$l + m + n \geqslant 1997 - l_1 + \sqrt{l_1(3994 - l_1)}.$$

WLOG assume $l \geqslant m$, then $l > 1400$, so $1 \leqslant l_1 \leqslant 60$.

Note that $1997 - l_1$ decreases by 1 each time l_1 increases by 1, but $\sqrt{l_1(3994 - l_1)}$ increases much faster, so we guess that the minimum will be achieved when l_1 is relatively small.

(1) $l_1 = 1$, $(m + n)(m - n) = 3993$, $m + n \geqslant 121$. Hence, when $l = 1996$, $m + n + l \geqslant 1996 + 121 = 2117$.

(2) When $200 \leqslant l_1 \leqslant 600$,

$$l + m + n \geqslant 1997 - 600 + \sqrt{200 \times 3794} > 2117.$$

(3) When $50 \leqslant l_1 \leqslant 200$, it's also easy to see that

$$l + m + n \geqslant 1997 - 200 + \sqrt{50 \times 3994} > 2117.$$

(4) When $10 \leqslant l_1 \leqslant 50$,

$$l+m+n \geqslant 1997-50+\sqrt{10\times 3984} > 2117.$$

(5) When $5 \leqslant l_1 \leqslant 10, l+m+n \geqslant 1997-10+\sqrt{5\times 3989} > 2117$.

(6) $l_1 = 4$, $l+m+n \geqslant 1997-4+2\times 63 > 2117$.

(7) $l_1 = 3$, $(m-n)(m+n) = 3\times 3991$. So $(m+n)_{\min} = 307$, and $l+m+n = 1997-3+307 = 2301$.

(8) $l_1 = 2$, $(m-n)(m+n) = 998$, then $l+m+n \geqslant 1997-2+998 > 2117$.

From the above discussion, $(l+m+n)_{\min} = 2117$, and the equality holds when $l = 1996$, $m = 77$, $n = 44$.

Next, we find the maximum value of $l+m+n$.

We have $(m-n)(m+n) = l_1(3994-l_1)$, hence $l_1 < m-n \leqslant m+n < 3994-l_1$, and that l_1, $m-n$, $m+n$, $3994-l_1$ are all even numbers or all odd numbers. Thus $m-n \geqslant l_1+2$, so $m+n \leqslant \dfrac{(3994-l_1)l_1}{l_1+2}$. We guess that the maximum is achieved when the equality holds.

Suppose the equality holds, then there may be two cases:

(i) l_1 is odd number, it suffices that $(l_1+2) \mid (3994-l_1)l_1$.

(ii) l_1 is even, then $(l_1+2) \mid (3994-l_1)l_1$, and the ratio must be even as well.

Denote $l' = l_1+2$, then $\dfrac{(3994-l_1)l_1}{l_1+2} = 3998-\left(l'+\dfrac{2\times 3996}{l'}\right)$.

So to make the equality hold, when l' is odd, we only need $l' \mid 2\times 3996$; when l' is even, we need $\dfrac{2\times 3996}{l'}$ to be even, so it suffices that $l' \mid 3996$. As a result,

$$l+m+n \leqslant 5997-2\left(l'+\frac{3996}{l'}\right),$$

and when $l' = 54$ or 74, $l+m+n \leqslant 5741$. We show in the following that this is indeed the maximum.

First, when l' is outside $[54, 74]$, $l+m+n$ clearly will not exceed 5741.

When $l' \in [54, 74]$, the equality cannot hold, therefore

$$l+m+n \leqslant 1997-l_1+\frac{l_1(3994-l_1)}{l_1+4} \leqslant 1997-l_1+\frac{71}{75}(3994-l_1)$$
$$=1997-l_1+(3994-l_1)-\frac{4}{75}(3994-l_1)$$
$$\leqslant 6000-100-\frac{4}{75}\times 3750<5741.$$

Therefore $(l+m+n)_{\max}=5741$, and the maximum is achieved when $l=1945$, $m=1925$, $n=1871$.

Exercise 9

1. We provide two proofs for this problem.

(1) For a fixed t, we can find at most t such that $a_i \leqslant t$. We next estimate the number of a_i's that are greater than t. First, for any a_i, a_j, we have $[a_i, a_j] \leqslant n$, in other words,

$$\frac{a_i a_j}{(a_i, a_j)} \leqslant n. \quad ①$$

But $(a_i, a_j) \mid a_i - a_j (\neq 0)$, thus $|a_i - a_j| \geqslant (a_i, a_j)$. From ①, we have $\frac{(a_i, a_j)}{a_i a_j} \geqslant \frac{1}{n}$, so $\frac{|a_i - a_j|}{a_i a_j} \geqslant \frac{1}{n}$, or $\left|\frac{1}{a_i}-\frac{1}{a_j}\right| \geqslant \frac{1}{n}$. In addition, for $a_1 > a_2 > \cdots > a_l > t$, we have $\frac{1}{n}<\frac{1}{a_1}<\cdots<\frac{1}{a_l}<\frac{1}{t}$. As a result,

$$l \leqslant \left(\frac{1}{t}-\frac{1}{n}\right)\cdot n,$$

therefore $k \leqslant t+\left(\frac{1}{t}-\frac{1}{n}\right)\cdot n=\frac{n}{t}+t-1$. Set $t=[\sqrt{n}]$, we have $k \leqslant 2\sqrt{n}+1$.

(2) Since $[a_1, a_2]=\frac{a_1 a_2}{(a_1, a_2)}$, we have $a_2 \cdot \left[\frac{a_1}{(a_1, a_2)}\right] \leqslant n$. As $a_1 > a_2$, it's easy to see that $\frac{a_1}{(a_1, a_2)} \geqslant 2$, thus $2a_2 \leqslant n$. With $a_1 \leqslant n$ and $2a_2 \leqslant n$, we guess that for all i, $1 \leqslant i \leqslant k$, we have

$$ia_i \leqslant n. \qquad ②$$

If ② is indeed true, then

$$\begin{aligned} k &= \text{number of } a_i\text{'s that is} \leqslant a_t + \text{number of } a_i\text{'s that is} > a_t \\ &\leqslant \text{number of } a_i\text{'s that is} \leqslant a_t + t \\ &\leqslant a_t + t \leqslant \frac{n}{t} + t. \end{aligned}$$

Setting $t = [\sqrt{n}] + 1$, we would then have $k \leqslant 2\sqrt{n} + 1$. So now it suffices to prove ②.

We turn to mathematical induction. Assume that $(i-1)a_{i-1} \leqslant n$ and we need to show that $ia_i \leqslant n$. In fact, $[a_i, a_{i-1}] = Aa_i = Ba_{i-1}$, and $[a_i, a_{i-1}] \leqslant n$.

(a) If $A \geqslant i$, then clearly $ia_i \leqslant n$.

(b) If $A < i$, since $a_i < a_{i-1}$, it follows that $B < A$.

$$ia_i = \frac{iB}{A}a_{i-1} = \frac{(i-1)iBa_{i-1}}{A(i-1)} \leqslant n\frac{iB}{(i-1)A}.$$

Since $A \geqslant B + 1$, we have

$$n\frac{iB}{(i-1)A} \leqslant n\frac{i(A-1)}{(i-1)A} < n\frac{i}{i-1}\left(1 - \frac{1}{i}\right) = n.$$

This completes the proof.

2. For part (1), denote the sum on the left hand side of the inequality as $f(x)$. Obviously when $x = 0$ or π, the inequality holds. Also, since $|f(x)|$ is an even function and periodic with period π, it suffices to prove the result for $x \in (0, \pi)$.

For any fixed $x \in (0, \pi)$, pick $m \in \mathbf{N}$, such that $m \leqslant \frac{\sqrt{\pi}}{x} < m + 1$.

Note that

$$\left|\sum_{k=1}^{n} \frac{\sin kx}{k}\right| \leqslant \left|\sum_{k=1}^{m} \frac{\sin kx}{k}\right| + \left|\sum_{k=m+1}^{n} \frac{\sin kx}{k}\right|,$$

where we understand the first summand of the RHS to be 0 when $m = 0$, and we understand the second summand to be 0 when $m \geqslant n$ and k

only goes up to n in the first summand.

Since $|\sin x| \leqslant |x|$, we have $\left|\sum_{k=1}^{m} \frac{\sin kx}{kx}\right| \leqslant \sum_{k=1}^{m} \frac{kx}{k} = mx \leqslant \sqrt{\pi}$. Further, we write $s_i = \sum_{k=m+1}^{i} \sin kx$ $(i = m+1, m+2, \cdots, n)$, and observe that

$$s_i \cdot \sin \frac{x}{2} = \frac{1}{2} \sum_{k=m+1}^{i} \left[\cos\left(k - \frac{1}{2}\right)x - \cos\left(k + \frac{1}{2}\right)x\right]$$
$$= \frac{1}{2}\left[\cos\left(m + \frac{1}{2}\right)x - \cos\left(i + \frac{1}{2}\right)x\right],$$

thus $|s_i| \leqslant 1/\sin \frac{x}{2}$ $(i = m+1, m+2, \cdots, n)$. Therefore,

$$m_1 = \max s_i \leqslant 1/\sin \frac{x}{2}, \; m_0 = \min s_i \geqslant -1/\sin \frac{x}{2}.$$

Denote $a_k = \sin kx$, $b_k = \frac{1}{k}$ $(k = m+1, m+2, \cdots, n)$, we have

$$b_{m+1} \geqslant b_{m+2} \geqslant \cdots \geqslant b_n.$$

According to Abel's Inequality,

$$m_0 b_{m+1} \leqslant \sum_{k=m+1}^{n} \frac{\sin kx}{k} \leqslant m_1 b_{m+1}.$$

Hence

$$\left|\sum_{k=m+1}^{n} \frac{\sin kx}{k}\right| \leqslant \frac{1}{m+1} \cdot \frac{1}{\sin \frac{x}{2}}.$$

Since $\sin x > \frac{2}{\pi} x$ for $x \in \left(0, \frac{\pi}{2}\right]$, we have $\sin \frac{x}{2} > \frac{2}{\pi} \cdot \frac{x}{2} = \frac{x}{\pi}$ for $x \in (0, \pi)$. As a result,

$$\left|\sum_{k=m+1}^{n} \frac{\sin kx}{k}\right| \leqslant \frac{\pi}{x} \cdot \frac{1}{m+1} \leqslant \frac{\pi}{x} \cdot \frac{x}{\sqrt{\pi}} = \sqrt{\pi}.$$

Therefore,

$$\left|\sum_{k=1}^{n}\frac{\sin kx}{k}\right| \leqslant \sqrt{\pi}+\sqrt{\pi}=2\sqrt{\pi}.$$

Comment. We can use the above method to prove the following problem:

Assume $\{a_n\}$ is a positive and non-increasing series, show that if for $n \geqslant 2001$, $na_n \leqslant 1$, then for any integer $m \geqslant 2001$, and $x \in \mathbf{R}$, we have

$$\left|\sum_{k=2001}^{m} a_k \sin kx\right| \leqslant 1+\pi.$$

We prove this in the following:

Let $f_{m,n}(x)=\sum_{k=n}^{m} a_k \sin kx$ $(n=2001)$. $f_{m,n}(x)$ is an odd function and periodic, so we only need to consider the interval $[0, \pi]$.

(i) $x \in \left[\frac{\pi}{n}, \pi\right]$, then

$$|\sin nx+\cdots+\sin mx| = \left|\frac{\cos\left(n-\frac{1}{2}\right)x-\cos\left(m+\frac{1}{2}\right)x}{2\sin\frac{x}{2}}\right|$$

$$\leqslant \frac{1}{\sin\frac{x}{2}} \leqslant \frac{\pi}{x}.$$

Thus $|f_{m,n}(x)| \leqslant a_n \cdot \frac{\pi}{x} \leqslant na_n \leqslant 1$.

(ii) $x \in \left[0, \frac{\pi}{m}\right]$, then

$$|f_{m,n}(x)| \leqslant a_n nx+\cdots+a_m mx \leqslant (m-n)x \leqslant \pi.$$

(iii) $x \in \left[\frac{\pi}{m}, \frac{\pi}{n}\right]$, so $n \leqslant \frac{\pi}{x} \leqslant m$. Denote $k=\left[\frac{\pi}{x}\right]$. Thus, $|f_{m,n}(x)| \leqslant |a_n \sin nx+\cdots+a_k \sin kx|+|a_{k+1}\sin(k+1)x+\cdots+a_m \sin mx|$.

From $0<x \leqslant \frac{\pi}{k}$ and (ii), we know $|a_n \sin nx+\cdots+a_k \sin kx| \leqslant \pi$.

From $\frac{\pi}{k+1} \leqslant x \leqslant \pi$ and (i), we know $|f_{m,k+1}(x)| \leqslant 1$. Therefore, $|f_{m,n}(x)| \leqslant |f_{k,n}(x)| + |f_{m,k+1}(x)| \leqslant 1 + \pi$.

For part (2) of the problem, use the result from problem 13 in exercise 2.

3. Set $y_i = \frac{x_{i+1}x_{i+2}}{x_i^2}$ $(i = 1, 2, \cdots, n, x_{n+1} = x_1, x_{n+2} = x_2)$. The original inequality is equivalent to

$$\frac{1}{1+y_1} + \frac{1}{1+y_2} + \cdots + \frac{1}{1+y_n} \leqslant n - 1. \qquad ①$$

From the definition of y_i, we have $y_1 y_2 \cdots y_n = 1$. If there exists two terms in ① that are $\leqslant \frac{1}{2}$, since $\frac{1}{1+y_i} < 1$, ① clearly holds.

If only $\frac{1}{1+y_1} \leqslant \frac{1}{2}$, then $y_1 \geqslant 1$, $y_2 \leqslant 1$, $y_3 \leqslant 1$, $\cdots$, $y_n \leqslant 1$. Hence,

$$\frac{1}{1+y_1} + \frac{1}{1+y_2} \leqslant \frac{1}{1+y_1} + \frac{1}{1+y_2y_3\cdots y_n} = \frac{1}{1+y_1} + \frac{1}{1+\frac{1}{y_1}} = 1.$$

Because the sum of the first two terms in ① is $\leqslant \frac{1}{2}$ while each of the other $n-2$ terms is <1, ① holds.

4. (1) Since

$$\begin{aligned}\left(\sum_{i=1}^{n} x_i\right)^2 &= \sum_{i=1}^{n} x_i^2 + 2\sum_{1 \leqslant i < j \leqslant n} x_i x_j \\ &= \sum_{i=1}^{n} x_i^2 + \sum_{1 \leqslant i < j \leqslant n} (x_i x_j)^2 + \sum_{1 \leqslant i < j \leqslant n} (2x_i x_j - (x_i x_j)^2) \\ &\leqslant \frac{n(n+1)}{2} + \sum_{1 \leqslant i < j \leqslant n} 1 = n^2,\end{aligned}$$

we have $\sum_{i=1}^{n} x_i \leqslant n$, and the equality holds when $x_1 = x_2 = \cdots = x_n = 1$.

(2) When $n = 1$, $x_1 = 1$, so $\sum_{i=1}^{n} x_i \geqslant \sqrt{\frac{n(n+1)}{2}}$.

When $n=2$, since $x_1^2+x_2^2+(x_1x_2)^2=3$, $x_1x_2\leqslant\sqrt{3}$, and

$(x_1+x_2)^2=3-x_1^2x_2^2+2x_1x_2=3+x_1x_2(2-x_1x_2)\geqslant 3.$

Thus $x_1+x_2\geqslant\sqrt{3}$, and the equality holds when $x_1=\sqrt{3}$, $x_2=0$.
When $n=3$, we are given $x_1^2+x_2^2+x_3^2+x_1^2x_2^2+x_2^2x_3^2+x_3^2x_1^2=6$.
If $\max\{x_1x_2, x_2x_3, x_3x_1\}\leqslant 2$, then

$$\begin{aligned}&(x_1+x_2+x_3)^2\\=&6+x_1x_2(2-x_1x_2)+x_2x_3(2-x_2x_3)+x_3x_1(2-x_3x_1)\geqslant 6.\end{aligned}$$

If $\max\{x_1x_2, x_2x_3, x_3x_1\}>2$, WLOG assume $x_1x_2>2$, then

$$x_1+x_2+x_3\geqslant x_1+x_2\geqslant 2\sqrt{x_1x_2}>2\sqrt{2}>\sqrt{6}.$$

When $n\geqslant 4$, set $x_1=x_2=x$, $x_3=x_4=\cdots=x_n=0$. By the given condition

$$2x^2+x^4=\frac{n(n+1)}{2},\text{ so }x=\pm\sqrt{\frac{\sqrt{2n(n+1)+4}-2}{2}}.$$

Omitting the negative root,

$$\sum_{i=1}^{n}x_i=\sqrt{2\sqrt{2n(n+1)+4}-4}<\sqrt{\frac{n(n+1)}{2}}.$$

From the above discussion, $n=1$, 2 or 3.

5. (1) When $a\geqslant 1\geqslant b\geqslant c$, $\theta=2+abc-ab-bc-ca$. Since $\frac{a}{c}+ac+\frac{b}{a}+ab+\frac{c}{b}+bc\geqslant 2a+2b+2c=6$. Also $a+b+c\geqslant 3\sqrt[3]{abc}$, $abc\leqslant 1$, so the LHS $\geqslant 6+abc-1-ab-bc-ca=3+\theta$.

(2) When $a\geqslant b\geqslant 1\geqslant c$, $\theta=ab+bc+ca-2-abc\geqslant 0$, hence $ab+bc+ca\geqslant 2+abc$. In addition, since

$$\frac{a}{c}+ab^2c\geqslant 2ab,\ \frac{b}{a}+abc^2\geqslant 2bc,\ \frac{c}{b}+a^2bc\geqslant 2ac,$$

$$\begin{aligned}\text{the LHS}&\geqslant 2(ab+bc+ca)-3abc\geqslant ab+bc+ca+2-2abc\\&\geqslant ab+bc+ca+1-abc=3+\theta.\end{aligned}$$

Comment. We can even prove a stronger result that

$$\frac{a}{c}+\frac{b}{a}+\frac{c}{b}\geqslant 3+4\theta.$$

Actually, since

$$\left(\frac{a}{c}+\frac{b}{a}+\frac{c}{b}\right)-\left(\frac{a}{b}+\frac{b}{c}+\frac{c}{a}\right)=\frac{(a-b)(b-c)}{bc}+\frac{(b-c)(b-a)}{ab}$$
$$=\frac{(a-b)(b-c)(a-c)}{abc}\geqslant 0,$$

we have

$$\frac{a}{c}+\frac{b}{a}+\frac{c}{b}-3\geqslant\frac{1}{2}\left(\frac{a}{c}+\frac{b}{a}+\frac{c}{b}+\frac{a}{b}+\frac{b}{c}+\frac{c}{a}\right)-3$$
$$=\frac{a(b-c)^2+b(c-a)^2+c(a-b)^2}{2abc}.$$

We divide our attention to the following two cases:

(i) $a\geqslant 1\geqslant b\geqslant c$. Assume $1-b=\alpha$, $1-c=\beta$, $\alpha\beta\geqslant 0$. Then $a-1=\alpha+\beta$, $\alpha+\beta<2$. Thus,

$$\frac{a(b-c)^2+b(c-a)^2+c(a-b)^2}{2abc}$$
$$=\frac{a(\alpha-\beta)^2+b(\alpha+2\beta)^2+c(\beta+2\alpha)^2}{2abc}$$
$$\geqslant\frac{b(\alpha+2\beta)^2+c(\beta+2\alpha)^2}{2abc}$$
$$\geqslant\frac{2\sqrt{bc}\cdot(\alpha+2\beta)(\beta+2\alpha)}{2abc}$$
$$=\frac{(\alpha+2\beta)(\beta+2\alpha)}{a\sqrt{bc}}\geqslant\frac{9\alpha\beta}{a\sqrt{bc}}=\frac{9\alpha\beta}{\sqrt{4\cdot\left(\frac{a}{2}\right)^2bc}}$$
$$\geqslant\frac{9\alpha\beta}{\sqrt{4\cdot\left(\frac{\frac{a}{2}+\frac{a}{2}+b+c}{4}\right)^4}}$$
$$=8\alpha\beta\geqslant 4(\alpha+\beta)\alpha\beta=4\theta.$$

(ii) $a\geqslant b\geqslant 1\geqslant c$. Assume $1+\alpha=a$, $1+\beta=b$, $c=1-\alpha-\beta$.

Then $\alpha+\beta\leqslant 1$, $\alpha-\beta\geqslant 0$. Therefore

$$\begin{aligned}&\frac{a(b-c)^2+b(c-a)^2+c(a-b)^2}{2abc}\\&=\frac{a(2\beta+\alpha)^2+b(2\alpha+\beta)^2+c(\alpha-\beta)^2}{2abc}\\&\geqslant\frac{a(2\beta+\alpha)^2+b(2\alpha+\beta)^2}{2abc}\geqslant\frac{2\sqrt{ab}(2\beta+\alpha)(2\alpha+\beta)}{2abc}\\&\geqslant 9\alpha\beta\geqslant 9(\alpha+\beta)\alpha\beta=9\theta.\end{aligned}$$

Combining (i) and (ii), we have shown $\frac{a}{c}+\frac{b}{a}+\frac{c}{b}\geqslant 3+4\theta$.

6. (1) When $a\leqslant 1$, $b\leqslant 1$, $c\leqslant 1$, we want to show LHS $\geqslant 3\geqslant a+b+c+\theta$, which is equivalent to $ab+bc+ca\leqslant 2+abc$.

Since $(1-a)(1-b)\geqslant 0$, we have $2-a-b\geqslant 1-ab\geqslant c(1-ab)$, it follows that

$$2+abc\geqslant a+b+c\geqslant ab+bc+ca.$$

(2) When $a\geqslant 1$, $b\leqslant 1$, $c\leqslant 1$, since $\theta\geqslant 0$, we have

$$a+b+c\geqslant ab+bc+ca+1-abc.$$

But LHS $+ac+ab+bc\geqslant 2(a+b+c)$, so

$$\text{LHS}\geqslant 2\sum a-\sum ac+abc-1=a+b+c+\theta.$$

(3) When $a\geqslant 1$, $b\geqslant 1$, $c\leqslant 1$, it's easy to show that $a+b+c\geqslant a+1+bc\geqslant 2+abc$.

Moreover, since $\theta\geqslant 0$, $ab+bc+ca\geqslant a+b+c+abc-1$.

$$\sum a^2b+\sum b\geqslant 2\sum ab\geqslant\sum ab+\sum a+abc-1,$$

this implies

$$\sum a^2b\geqslant\sum ab+abc-1\geqslant abc\left(\sum ab+1-abc\right).$$

Therefore, LHS $\geqslant\sum ab+1-abc=a+b+c+\theta$. We can show the result similarly when $a\geqslant 1$, $c\geqslant 1$, $b\leqslant 1$.

7. (1) We show that, when we order the six circles with radius 1, 3, 5, 6, 4, 2 or 2, 4, 6, 5, 3, 1 accordingly, the common tangent line segment between the leftmost and rightmost circles reaches its maximum.

Assume the six circles in order are O_1, O_2, $\cdots$, O_6 and the radius of O_i is r_i. These six circles are tangent to l at points $T_i(1 \leqslant i \leqslant 6)$ respectively. For $1 \leqslant i \leqslant 5$, by Pythagoras' theorem we have $T_iT_{i+1}^2 = 4r_ir_{i+1}$. Thus,

$$T_1T_6 = 2(\sqrt{r_1r_2} + \sqrt{r_2r_3} + \sqrt{r_3r_4} + \sqrt{r_4r_5} + \sqrt{r_5r_6}).$$

Because r_1, r_2, $\cdots$, r_6 is a permutation of 1, 2, $\cdots$, 6, T_1T_6 can be expressed alternatively as

$$T_1T_6 = 2(\sqrt{1r'_1} + \sqrt{2r'_2} + \sqrt{3r'_3} + \sqrt{4r'_4} + \sqrt{5r'_5}).$$

From the above and the Rearrangement Inequality, we know that T_1T_6 reaches its maximum when $r'_1 \leqslant r'_2 \leqslant r'_3 \leqslant r'_4 \leqslant r'_5$ and each of these is as large as possible. In this light, we try to set $r'_5 = r'_4 = 6$ and see if we can find suitable values for r'_1, r'_2 and r'_3.

Since r'_1, r'_2, r'_3 can only take distinct values from 1, 2, 3, 4, 5, we set $r'_1 = 3$, $r'_2 = 4$, $r'_3 = 5$. Such selection of r'_1, r'_2, $\cdots$, r'_6 will certainly maximize T_1T_6. But is it possible? Note that

$$\begin{aligned}\sum_{i=1}^{5} \sqrt{ir'_i} &= \sqrt{1\times 3} + \sqrt{2\times 4} + \sqrt{3\times 5} + \sqrt{4\times 6} + \sqrt{5\times 6}\\ &= \sqrt{1\times 3} + \sqrt{3\times 5} + \sqrt{5\times 6} + \sqrt{6\times 4} + \sqrt{4\times 2}\\ &= \sqrt{2\times 4} + \sqrt{4\times 6} + \sqrt{6\times 5} + \sqrt{5\times 3} + \sqrt{3\times 1}.\end{aligned}$$

So indeed we can find proper arrangement of the six circles so that T_1T_6 reaches its maximum as determined above.

(2) Next, we show that when we order the six circles with radius 6, 1, 4, 3, 2, 5 or 5, 2, 3, 4, 1, 6 accordingly, $M = T_1T_6$ reaches its minimum value. If we arrange r_1, $\cdots$, r_6 on a circle,

$$T_1T_6 = 2([\sum_{i=1}^{6} \sqrt{r_ir_{i+1}}] - \sqrt{r_kr_{k+1}})$$

when we break the circular arrangement at r_k. Because $1\times 6 \leqslant 2\times 3$, it's easy to see that we should not break at 1, or in other words, we should not put the circle with radius 1 at the two sides. Due to symmetry we only need to consider the circle with radius 1 being placed at the second or the third position.

When $\{r_2, r_3, \cdots, r_6\} = \{2, 3, \cdots, 6\}$, the arrangement $r_2, r_3, 1, r_4, r_5, r_6$ gives $T_1T_6 = \sqrt{r_2r_3} + \sqrt{r_3} + \sqrt{r_4} + \sqrt{r_4r_5} + \sqrt{r_5r_6}$. Assume $\{a, b\} \subset \{2, 3, 4, 5, 6\}$ and $a < b$.

(i) If r_4, r_5, r_6 are fixed, it's easy to see that T_1T_6 is smaller when $r_2 > r_3$.

(ii) If r_2, r_3, r_6 are fixed, assume r_4, r_5 takes value from $\{a, b\}$, and the relative difference of T_1T_6 is given by $\sqrt{a} + \sqrt{br_6} - \sqrt{b} - \sqrt{ar_6} = (\sqrt{b} - \sqrt{a})(\sqrt{r_6} - 1) > 0$. Thus T_1T_6 is smaller when $r_4 > r_5$. Now, we assume r_4, r_5, r_6 takes value from $\{a, b, c\}$, and that $a < b < c$.

Since $\sqrt{c} + \sqrt{cb} + \sqrt{ba} > \sqrt{c} + \sqrt{ca} + \sqrt{ab} > \sqrt{b} + \sqrt{ba} + \sqrt{ac}$, it turns out when $r_6 > r_4 > r_5$, T_1T_6 is smaller.

(iii) If r_3, r_4, r_5 are fixed, assume r_2, r_6 takes value from $\{a, b\}$, the relative difference of T_1T_6 is given by $\sqrt{ar_3} + \sqrt{r_5b} - \sqrt{br_3} - \sqrt{r_5a} = (\sqrt{b} - \sqrt{a})(\sqrt{r_5} - \sqrt{r_3})$. Thus when $r_5 > r_3$ and $r_2 > r_6$, or when $r_5 < r_3$ and $r_2 < r_6$, T_1T_6 is smaller.

From the above discussion, when $r_2 > r_6 > r_4 > r_5 > r_3$ or when $r_6 > r_4 > r_5$ and $r_6 > r_2 > r_3 > r_5$, T_1T_6 is smaller. They reflect the following arrangements: (a) 6, 2, 1, 4, 3, 5; (b) 4, 3, 1, 5, 2, 6; (c) 5, 3, 1, 4, 2, 6; (d) 5, 4, 1, 3, 2, 6.

Correspondingly, the values of T_1T_6 are given by $M_1 = 4\sqrt{3} + \sqrt{2} + 2 + \sqrt{15}$, $M_2 = 5\sqrt{3} + \sqrt{5} + \sqrt{10}$, $M_3 = \sqrt{15} + 3\sqrt{3} + 2 + 2\sqrt{2}$, $M_4 = 2\sqrt{5} + 2 + \sqrt{6} + 3\sqrt{3}$. M_3 gives the minimum of T_1T_6 here, with arrangement 5, 3, 1, 4, 2, 6.

Next, we find the minimum value of T_1T_6 when the arrangement looks like $r_2, 1, r_3, r_4, r_5, r_6$. Now

$$T_1T_6 = \sqrt{r_2} + \sqrt{r_3} + \sqrt{r_3r_4} + \sqrt{r_4r_5} + \sqrt{r_5r_6},$$

and similarly we can show that when $r_2 > r_6 > r_3 > r_4 > r_5$ or when $r_2 > r_3 > r_6 > r_5 > r_4$, T_1T_6 is smaller. These orderings give the following arrangements: (a) 6, 1, 4, 3, 2, 5; (b) 6, 1, 5, 2, 3, 4.

Correspondingly, T_1T_6 achieve values $M_1 = 2\sqrt{6} + 2 + 2\sqrt{3} + \sqrt{10} < M_2 = 2\sqrt{6} + \sqrt{5} + 2\sqrt{3} + \sqrt{10}$.

Combining all the cases above, T_1T_6 is minimized with arrangement 6, 1, 4, 3, 2, 5 or 5, 2, 3, 4, 1, 6.

8. Assume the five numbers are $0 < x_1 \leqslant x_2 \leqslant x_3 \leqslant x_4 \leqslant x_5$. First we have $x_4 < \frac{1}{2}$, otherwise $x_4 \geqslant \frac{1}{2}$, $x_5 \geqslant \frac{1}{2}$, then $x_4 + x_5 \geqslant 1$ which leads to a contradiction. Hence $x_5 = \frac{1}{2}$.

Pick x_1, x_2, there exists $x_i\,(3 \leqslant i \leqslant 5)$, such that $x_1 + x_2 + x_i = 1$.

Pick x_4, x_5, there exists $x_j\,(1 \leqslant j \leqslant 3)$, such that $x_4 + x_5 + x_j = 1$.

Thus, $1 = x_1 + x_2 + x_i \leqslant x_1 + x_3 + x_i \leqslant x_1 + x_4 + x_i \leqslant x_1 + x_4 + x_5 \leqslant x_j + x_4 + x_5 = 1$.

This implies $x_1 + x_2 + x_i = x_1 + x_3 + x_i = x_1 + x_4 + x_i$, so $x_2 = x_3 = x_4 = x$.

As a result we have three possible cases:

(1) $x_1 + \frac{1}{2} + x = 1$; $x_1 + x + x = 1$. (2) $x_1 + \frac{1}{2} + x = 1$; $2x + x = 1$. (3) $x_1 + \frac{1}{2} + x = 1$; $2x + \frac{1}{2} = 1$.

There is no solution for (1), (2) gives $x = \frac{1}{3}$, $x_1 = \frac{1}{6}$ and (3) gives $x = x_1 = \frac{1}{4}$.

Therefore, $(x_1, x_2, x_3, x_4, x_5) = \left(\frac{1}{2}, \frac{1}{3}, \frac{1}{3}, \frac{1}{3}, \frac{1}{6}\right)$ or $\left(\frac{1}{2}, \frac{1}{4}, \frac{1}{4}, \frac{1}{4}, \frac{1}{4}\right)$.

9. WLOG assume $x \geqslant y \geqslant z$. Thus,

$$\begin{aligned}
&x^2y^2 + y^2z^2 + z^2x^2 + x^2y^2z^2 \\
&= x^2 \cdot [(1-x)^2 - 2yz] + y^2z^2 + x^2y^2z^2
\end{aligned}$$

$$= x^2(1-x)^2 - 2x^2yz + y^2z^2 + x^2y^2z^2$$
$$\leqslant \frac{1}{16} - (x^2yz - y^2z^2) - (x^2yz - x^2y^2z^2) \leqslant \frac{1}{16}.$$

The equality holds when $x = y = \frac{1}{2}$, $z = 0$.

10. First $x^2y + y^2z + z^2x \geqslant 0$, when $x = 1$, $y = z = 0$ the equality holds, so the minimum value is 0. On the other hand, WLOG assume $x = \max\{x, y, z\}$, then (1) If $x \geqslant y \geqslant z$, we have $x^2y + y^2z + z^2x \leqslant x^2y + y^2z + z^2x + z[xy + (x-y)(y-z)] = (x+z)^2y = (1-y)^2y \leqslant \frac{4}{27}$.

(2) If $x \geqslant z \geqslant y$, since $(x-y)(y-z)(z-x) = (xy^2 + yz^2 + zx^2) - (x^2y + y^2z + z^2x)$, $x^2y + y^2z + z^2x \leqslant x^2y + y^2z + z^2x + (x-y)(y-z)(z-x) = x^2z + z^2y + y^2x \leqslant \frac{4}{27}$. Therefore, the maximum value is $\frac{4}{27}$, and is achieved with $x = \frac{2}{3}$, $y = \frac{1}{3}$, $z = 0$.

11. WLOG assume $a \geqslant b \geqslant c$. Let $b = c + x$, $a = b + y = c + x + y$. Since $a < b + c$, $y < c$. If $xy = 0$ the statement is obviously true. Next we assume $xy \neq 0$, then $x > 0$, $y > 0$.

$$\begin{aligned}\text{LHS} &= \frac{yx(x+y)}{(2x+y+2c)(x+2c)(x+y+2c)} \\ &< \frac{x^2y + y^2x}{(2x+3y)(x+2y)(x+3y)} \\ &= \frac{x^2y + y^2x}{2x^3 + 13x^2y + 27xy^2 + 18y^3}.\end{aligned}$$

Write $x = ky$, $k > 0$, then

$$\begin{aligned}&2x^3 + 13x^2y + 27xy^2 + 18y^3 \\ &= 2x^2 \cdot ky + 13x^2y + 22xy^2 + \frac{5}{k}x^2y + \frac{18}{k^2}x^2y \\ &= \left(2k + \frac{5}{k} + \frac{18}{k^2}\right)x^2y + 13x^2y + 22xy^2 \\ &\geqslant 5 \cdot \sqrt[5]{\left(\frac{2}{3}k\right)^3 \cdot \frac{5}{k} \cdot \frac{18}{k^2}} \cdot x^2y + 13x^2y + 22xy^2 \\ &> 22(x^2y + y^2x).\end{aligned}$$

So the original inequality holds.

12. Set $a = b = 2c$, we have $k \leqslant 100$. Since

$$\frac{a+b+c}{abc} \cdot [(a+b)^2 + (a+b+4c)^2]$$

$$\geqslant \frac{a+b+c}{abc} \cdot [(a+b)^2 + (a+2c+b+2c)^2]$$

$$\geqslant \frac{a+b+c}{abc} \cdot [4ab + (2\sqrt{2ac} + 2\sqrt{2bc})^2]$$

$$= \frac{a+b+c}{abc} \cdot (4ab + 8ac + 8bc + 16\sqrt{ac} \cdot c)$$

$$= (a+b+c)\left(\frac{4}{c} + \frac{8}{a} + \frac{8}{b} + \frac{16}{\sqrt{ab}}\right)$$

$$= \left(\frac{a}{2} + \frac{a}{2} + \frac{b}{2} + \frac{b}{2} + c\right) \cdot \left(\frac{4}{c} + \frac{8}{a} + \frac{8}{b} + \frac{8}{\sqrt{ab}} + \frac{8}{\sqrt{ab}}\right)$$

$$\geqslant \left(5\sqrt[5]{\frac{a^2b^2c}{2^4}}\right) \cdot \left(5\sqrt[5]{\frac{2^{14}}{a^2b^2c}}\right) = 100.$$

Therefore $k_{\max} = 100$.

13. (1) $p^2 + q^2 + 1 > p(q+1)$ is equivalent to $\left(q - \frac{p}{2}\right)^2 + \left(\frac{p}{2} - 1\right)^2 + \frac{p^2}{2} > 0$, which is obviously true.

(2) Set $p = \sqrt{2}$, $q = 1$, then $b \leqslant \sqrt{2}$. We prove in the following that $p^2 + q^2 + 1 \geqslant \sqrt{2}p(q+1)$ holds for all real numbers p, q. In fact, this is because $\left[\frac{p}{\sqrt{2}} - q\right]^2 + \left[\frac{p}{\sqrt{2}} - 1\right]^2 \geqslant 0$, So $b_{\max} = \sqrt{2}$.

(3) Set $p = q = 1$, we have $c \leqslant \frac{3}{2}$. We next show that $p^2 + q^2 + 1 \geqslant \frac{3}{2}p(q+1)$ holds for all integers p, q. In fact, this is equivalent to $(3p - 4q)^2 + (7p^2 - 24p + 16) \geqslant 0$.

For $p \geqslant 3$ or $p \leqslant 0$, $7p^2 - 24p + 16 \geqslant 0$ so the inequality holds. Also, for $p = 1, 2$, it's easy to check that the inequality holds. Therefore $c_{\max} = \frac{3}{2}$.

14. (1) WLOG assume $a \leqslant b \leqslant c$, thus $a + b > c$. The original inequality is equivalent to

$$\sqrt[n]{a^n + b^n} + \sqrt[n]{b^n + c^n} + \sqrt[n]{c^n + a^n} < (a + b + c) + \frac{\sqrt[n]{2}}{2}(a + b + c).$$

Note that $\frac{\sqrt[n]{2}}{2}(a + b + c) > \sqrt[n]{2} \cdot c = \sqrt[n]{c^n + c^n} \geqslant \sqrt[n]{b^n + c^n}$. In addition, it's not hard to show that $\sqrt[n]{a^n + b^n} \leqslant \frac{a}{2} + b$ and $\sqrt[n]{c^n + a^n} \leqslant \frac{a}{2} + c$. Adding these together yields the desired result.

(2) Set $a = b$, $c \to 0$, then LHS $\to \frac{2 + \sqrt[3]{2}}{4}$, so $k \geqslant \frac{2 + \sqrt[3]{2}}{4}$. Next, we show that LHS $< \frac{2 + \sqrt[3]{2}}{4}$.

First, note that if a, b, c are the three sides of a triangle, then $\sqrt{a}$, $\sqrt{b}$, $\sqrt{c}$ are also three sides of a triangle. WLOG assume $\sqrt{a} + \sqrt{b} + \sqrt{c} = 2$, and $x = \sqrt{a}$, $y = \sqrt{b}$, $z = \sqrt{c}$, thus $x + y + z = 2$, and x, y, $z \in (0, 1)$. We show in the following that

$$\sqrt[3]{x^6 + y^6} + \sqrt[3]{y^6 + z^6} + \sqrt[3]{z^6 + x^6} < 2 + \sqrt[3]{2}. \qquad ①$$

Since x, y, $z < 1$, the LHS of ① is $< \sqrt[3]{x^3 + y^3} + \sqrt[3]{y^3 + z^3} + \sqrt[3]{z^3 + x^3}$. But when $x \geqslant y > 0$, it's easy to see that $x^3 + y^3 \leqslant [x + (\sqrt[3]{2} - 1)y]^3$.

Now let's assume $x \geqslant y \geqslant z$, then in ①,

$$\begin{aligned}
\text{LHS} &\leqslant x + (\sqrt[3]{2} - 1)y + y + (\sqrt[3]{2} - 1)z + x + (\sqrt[3]{2} - 1)z \\
&= 2x + \sqrt[3]{2}y + 2(\sqrt[3]{2} - 1)z \text{ (we would like } x,\ y \to 1,\ z \to 0) \\
&= 2x + \sqrt[3]{2}y + (2\sqrt[3]{2} - 2)(2 - x - y) \\
&= 4\sqrt[3]{2} - 4 + (4 - 2\sqrt[3]{2})x + (2 - \sqrt[3]{2})y \\
&\leqslant 4\sqrt[3]{2} - 4 + 4 - 2\sqrt[3]{2} + 2 - \sqrt[3]{2} \\
&= 2 + \sqrt[3]{2}.
\end{aligned}$$

Therefore ① holds.

Printed in the United States
By Bookmasters